CONTENTS

FOREWORD

A new publications series for the
Quaternary Research Association

Quaternary Proceedings

This is the first issue of a new series of publications sponsored by the *Quaternary Research Association* reporting the proceedings of meetings held within the aegis of the *QRA* or organised jointly by the *QRA* and other scientific organisations. The aim of the *Quaternary Proceedings* series is to enable the rapid publication of important scientific information at prices the ordinary *QRA* members can afford. This is achievable by publication 'in-house' using the new generation of PC hardware and software that is now widely available. This first volume was produced by the staff of the City Cartographic and Desk-Top Publishing Unit, City of London Polytechnic, using an Apple Macintosh system and output on a high resolution Imagesetter.

Each volume of *Quaternary Proceedings* will be produced independently by an appointed editor or editorial team. Proposals for new volumes should be directed to the *Series Editor*. Enquiries concerning the sale of *Quaternary Proceedings* should be directed to John Wiley & Sons, Chichester.

PREFACE

This volume reports the proceedings of a one-day discussion meeting held at the Geological Society, Piccadilly, London on 7th February, 1990. The meeting was sponsored by the *Quaternary Research Association* and the *Geological Society* as one of the first formal discussion meetings of the *Joint Association for Quaternary Research*. The full programme of contributions at the meeting is represented here.

As organiser of the meeting, I would like to take the opportunity of thanking the following who helped to make it such a successful and enjoyable occasion. Professors Jim Rose (RHBNC, University of London) and Peter Worsley (Reading University) chaired the two main sessions while Dr. Don Sutherland (Edinburgh) led the final discussion session. These three colleagues, together with Dr. Doug Harkness (NERC Radiocarbon Laboratory, East Kilbride) also provided much advice and encouragement during the organisation and running of the meeting. All of the papers have been peer-reviewed by at least two referees. The referees were immensely helpful in giving their time freely to deal with the papers in a very short time. The following post-graduate students and technical staff of the Department of Geography, RHBNC assisted in the organisation and administration of the meeting: Vanessa Norwood, Glynis Read, Jerry Lee and Simon Lewis. Finally, the staff of the City Cartographic and Desk-Top Publishing Unit of the City of London Polytechnic, under the direction of Don Shewan, and Kathy Roberts, Secretary of the Department of Geography, RHBNC, have helped to expedite publication of the proceedings. I am grateful to everyone concerned.

J. John Lowe

Centre for Quaternary Research
Department of Geography
Royal Holloway
University of London
Egham, Surrey, TW20 0EX, UK

March, 1991

INTRODUCTION

Not so very long ago most Quaternary stratigraphers employing the radiocarbon method as a basis for chronology were content to obtain a set of dates demonstrating stratigraphic continuity and relatively small error ranges. If the dates approximated an already-accepted chronostratigraphic scheme, they would most often be as regarded as 'reliable'. Today it is becoming more and more evident that establishing the reliability of dates is not quite so straightforward. It requires a detailed knowledge of, for example, the stratigraphic integrity and geochemistry of the dated samples, calibration of samples to take into account the possible effects of temporal variations in atmospheric radiocarbon activity, and possible systematic inter-laboratory variations in measuring radiocarbon activity. The true margin of error associated with radiocarbon dates is often difficult to determine, but is likely to be far larger than most of us have assumed. Those not already convinced of this are recommended to read the paper by **Jon Pilcher** published in this volume.

The papers presented here illustrate well the range of problems that Quaternary stratigraphers need to take into account in order to establish the reliability of radiocarbon dates. This is especially important when dealing with sediment samples with true ages that lie beyond the present range of dendrochronological calibration. Individually and collectively, the papers have one major message: the production of ever greater numbers of radiocarbon dates obtained by 'conventional' approaches is not going to provide a reliable basis for chronology. Rather, it is imperative that we look more constructively at the way in which radiocarbon activity is measured, the stratigraphic and chemical context of the samples employed for dating, and the treatment of samples at all stages of investigation. The contents of this volume illustrate some of the ways by which the problems are being more clearly defined and solutions to some of them are being sought.

The papers fall into two groups. The first five papers deal with theoretical aspects of measuring and interpreting radiocarbon 'numbers'. **Marian Scott** and her associates report on the very important quality assurance exercise that the radiocarbon laboratories have embarked upon in order to establish a protocol for inter-laboratory checks and procedures. **Robert Hedges** provides a review of the potential advantages and current 'state-of-play' of the use of AMS methods in radiocarbon measurement. In stark contrast to the resource-intensive AMS method, **Claudio Vita-Finzi** exhorts us to consider the practical advantages of employing 'first-order ^{14}C-dating' in certain stratigraphic contexts, where the availability of inexpensive, rapidly-obtained results is considered to outweigh some of the technical disadvantages. **The editor** illustrates some of the difficulties of dating sediments of the last glacial-interglacial transition, especially where samples have been obtained from sequences of poor stratigraphic resolution. Finally in this group, **Jon Pilcher** offers some quite sobering thoughts on the likely error ranges associated with dates obtained by routine laboratory procedures, and makes some recommendations to ^{14}C users who wish to achieve much greater accuracy.

The second group of papers illustrate applications of radiocarbon measurements to solve particular stratigraphic problems. **Doug Harkness** and **Mike Walker** examine the value of selecting different chemical 'fractions' for ^{14}C measurement in assessing the ^{14}C chronology of Lateglacial sediments. **Richard Preece** compares a number of accelerator and radiometric dates obtained from both sediments and macrofossil material from colluvial deposits and finds a reassuring measure of consistency. **Svante Björck** and his associates report on a series of sometimes major errors associated with terrestrial, lacustrine and marine samples from the Antarctic Peninsula; they review some of the processes that may have produced the very heterogeneous collection of dates obtained in their recent work. **Edouard Bard** and his associates demonstrate how AMS ^{14}C dating is being used to examine rates of change in $\delta^{18}O$ signals observed in ocean cores and how these may be used to test models of meltwater discharge in the oceans. Finally, **Ken Creer** reports on a sequence of late Quaternary deposits from Lac du Bouchet, France, where a comparison of radiocarbon, palynological, palaeomagnetic and oxygen isotope data can be used to establish dating control for sediments that accumulated through the last glacial cycle.

The contents of the volume are quite diverse and draw attention to such a range of problems and doubts that readers may well question the wisdom of using radiocarbon dates at all ! That would be an unfortunate attitude to adopt. One great strength of the method is that it has attracted the attention of many scientists with widely varying backgrounds and experience which has encouraged a multi-disciplinary approach to interpretation of results. In trying to solve chronological problems, we have gained a much greater knowledge of many inter-related environmental processes, such as temporal variations in atmospheric radiocarbon levels, exchanges of radiocarbon between different C reservoirs and organisms, variations in stable C ratios and their relationships to climate and the chemical 'behaviour' of sediments and groundwater. Radiocarbon dating will continue to have extensive applications in Quaternary and environmental sciences for many years to come, and it is therefore vital that the integrity of samples and the reliability of activity measures remain subject to the sort of assiduous scrutiny that we have come to expect from the 'radiocarbon community'. It is the editor's belief that this volume makes a significant contribution to that tradition.

J. John Lowe

Centre for Quaternary Research
Department of Geography
Royal Holloway
University of London
Egham, Surrey, TW20 0EX, UK

March, 1991

Quaternary Proceedings No. 1, 1991 1-4
© Quaternary Research Association, Cambridge

Future Quality Assurance in ^{14}C Dating

E. Marian Scott, Doug D. Harkness, Gordon T. Cook, Tom C. Aitchison, Murdoch S. Baxter

E.Marian Scott, Doug D. Harkness, Gordon T. Cook, Tom C. Aitchison, Murdoch S. Baxter, 1991 Future Quality Assurance in ^{14}C Dating, in *Radiocarbon Dating: Recent Applications and Future Potential* (ed. J.J. Lowe). Quaternary Proceedings No. 1, John Wiley & Sons Ltd, Chichester, pp. 1-4.

Abstract

The ^{14}C community has recently agreed to undertake a quality assurance scheme designed to ensure reliability and comparability of results across laboratories. We present evidence which demonstrates the need for this action, and outline the remedial action intended by concerned dating laboratories.

Historically, ^{14}C labs have inter-calibrated irregularly with restricted numbers of samples to help ensure satisfactory performance. Large scale organised inter-lab comparisons have, however, revealed significant discrepancies amongst ^{14}C laboratories in the past. Recently, the second International Inter-Comparison (ICS)(Scott *et al.* 1990a) further demonstrated the degree and magnitude of such discrepancies. The ICS involved over fifty laboratories worldwide (including representatives of gas counting, liquid scintillation and accelerator laboratories) and was a three-stage project carried out over three years. Analysis of the results demonstrated the existence of systematic laboratory biases and additional sources of variability not accounted for by the quoted errors.

As a result of the study and following discussion of its findings at an international workshop (Scott, Long and Kra, 1990) proposals for quality control and assurance have been devised. These include the introduction of a published protocol for internal laboratory procedures; the introduction of additional reference materials, provided by the International Atomic Energy Agency in Vienna, and finally a provision for regular international intercomparisons to be organised by the present authors.

In essence, ^{14}C laboratories are entering a new era of internal and external accountability, and it is appropriate that the wider scientific community is aware of these developments.

KEYWORDS: Radiocarbon dating; quality assurance

E Marian Scott, Tom C. Aitchison, Department of Statistics, Glasgow University, Glasgow G12 8QQ

Doug D. Harkness, NERC Radiocarbon Laboratory, East Kilbride, Glasgow, G75 0QU

Gordon T. Cook, Murdoch S. Baxter, Glasgow University Radiocarbon Laboratory, East Kilbride, Glasgow G75 0QU

Introduction

The questions of the reliability and reproducibility of routinely acquired ^{14}C dates have been of concern to users for some time. Difficulties in equating results from different laboratories have been demonstrated (Waterbolk, 1983). Equally, ^{14}C laboratories have been concerned with the accuracy and precision of the dates which they produce. These concerns, although perhaps viewed somewhat differently by the two communities, are clearly related and have resulted in the development of proposals for quality assurance which are designed with the re-assurance of users as a primary goal.

The procedures of ^{14}C dating are complex and some safeguards to ensure reliability of results are already inbuilt. In the main, these analytical controls are dependent on calibration, through secondary and tertiary standards, to the primary laboratory reference material of oxalic acid (the activity of which is related to that in 1890 wood and its contemporary atmosphere). In addition, most laboratories have tended to intercalibrate on an irregular basis using restricted subsets of samples. Recent large-scale organised inter-comparisons have, however, tended to demonstrate the inadequacy of this ad-hoc approach. This paper summarises the evidence that has

confirmed certain shortcomings in ^{14}C dating quality control and provides a brief discussion of the consequent implications. These findings have now been adopted by the ^{14}C community in general and wide-ranging proposals for quality control and assurance are being developed at high priority. Details of these proposals will be described along with the rationale and logic behind the incorporation of each stage in what is a multi-stage procedure.

Implementation of the new plan is already underway demonstrating the considerable commitment on the part of the ^{14}C community towards improving and maintaining the quality of its routine results. Because these changes are, in essence, introducing a new era in radiocarbon dating, it is appropriate at this early stage to summarise the fundamental innovations.

International Collaborative Study

The most recent large-scale intercomparison of ^{14}C dating was completed at the end of 1989 (Scott *et al.* 1990b). Laboratories were directly and indirectly informed of the study (Scott *et al.* 1986) which lasted three years and involved participation by over fifty laboratories worldwide, including exponents of gas counting (GC), liquid scintillation (LS) and accelerator methods

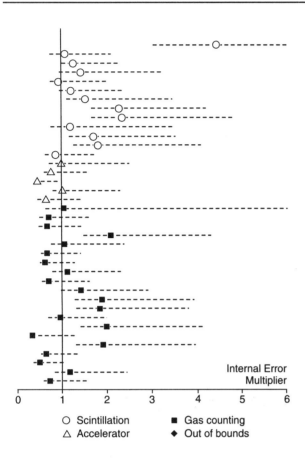

Figure 1 Internal error multiplier and 95% confidence interval.

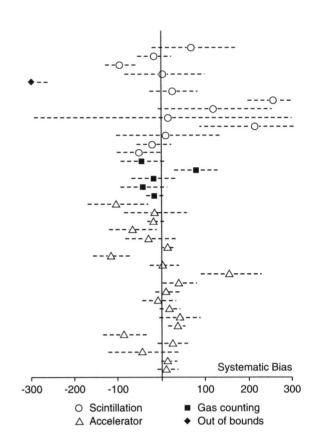

Figure 2 Systematic bias and 95% confidence interval.

(AMS). The full proceedings of an international workshop on the study appear in *Radiocarbon* Vol 32, 1990. It is, however, appropriate in the context of considering future developments in Quaternary reconstruction to review the purpose and design of this most recent inter-calibration exercise and to highlight its immediate implications for ¹⁴C laboratories and their customers or collaborators.

Experimental design.

The entire three year effort by participating laboratories was deliberately designed around a three-stage hierarchical framework. The initial stage concentrated on the counting processes, the second introduced the chemical conversion of sample carbon for subsequent counting and finally, a suite of typical 'raw material' samples was issued (Cook *et al.* 1990). Each component stage included unidentified duplicate samples, to allow assessment of internal consistency relative to the quoted laboratory errors. The test materials ranged in age from modern to *ca* 7000 years BP and comprised calcium carbonate and benzene (Stage 1), humic acid, algal lithothamnion and wood cellulose (Stage 2) and for the final total analyses (Stage 3) wood, marine shell and peat. Closing dates were enforced for the acceptance of data from each stage and the intercomparison was, as best as could be ensured, totally blind for all participants.

Statistical analysis.

Analysis of the results returned by the participating laboratories addressed three key concerns:

1) the role of the quoted error as a measure of internal consistency as indicated by the duplicate analyses

2) the existence, or otherwise, of systematic biases and the role of the quoted error in adequately explaining any such inter-laboratory variation

and

3) a comparison of the relative performance by different laboratory techniques (*i.e.* liquid scintillation(LS), gas counting(GC) and accelerator mass spectrometry(AMS)).

Three measures of laboratory performance were defined, two being directly linked to quoted error through the concept of an error multiplier. They are :

1) the Internal Error Multiplier (IEM) which is based on the differences between duplicate measurements and is used as a measure of internal precision: *i.e.* if the observed differences between duplicate samples are within the errors of the differences (the IEM is plausibly 1) then a lab is said to be internally consistent

2) the Bias which quantifies systematic offset from the true age (in this case taken to be the 'middle' value of all the results)

and

3) the External Error Multiplier (EEM) which relates quoted error to external variation after accounting for laboratory bias.

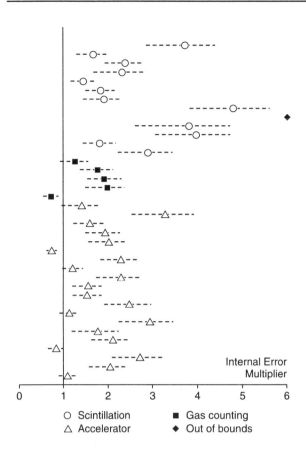

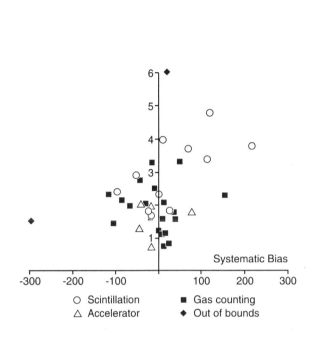

Figure 3 External error multiplier and 95% confidence interval.

Figure 4 Scatterplot of bias against external error multiplier.

The latter two indices are used to assess external consistency: i.e. a lab is said to be externally consistent if it is unbiased and its quoted error is adequate (the EEM is plausibly 1).

The Analysis and its Interpretation

Internal consistency

A plausible range of values for the IEM (formally a 95% confidence interval) was evaluated for each laboratory. Figure 1 indicates each interval, with the different laboratory types (GC, LS and AMS) being clearly indicated. A total of five gas counting plus six liquid scintillation laboratories had intervals wholly exceeding 1, suggesting that individually their quoted errors were too small.

External Consistency

Evaluation of systematic bias was carried out relative to a baseline defined by all the study results. Figure 2 shows the range of bias for each laboratory. Overall, five liquid scintillation, one AMS and six gas counting laboratories had significant biases of up to several hundred years (i.e. the range did not include zero). Some laboratories showed evidence that bias was changing on a relatively short term basis (from stage to stage).

At the same time, an EEM was evaluated for each laboratory. Figure 3 shows the resultant intervals for EEM. It should be pointed out that for those laboratories in which the bias at each stage changed considerably (either in sign or magnitude), then the external error multiplier will be amplified

as a result. Nevertheless, no liquid scintillation laboratory had an EEM which could be 1; only two AMS labs and six gas counting labs had EEMs which could plausibly be 1, indicating that their quoted errors were adequate.

Figure 4 shows a scatterplot of EEM against systematic bias for all laboratories, with the laboratory type indicated. It is clear from this diagram that a small group of laboratories has satisfactorily small biases and adequate quoted errors: none of these laboratories uses liquid scintillation counting.

Overall Performance

We have defined three criteria of acceptability, namely no significant systematic bias and adequate assessment of internal and external variability. The major mode of failure in the study has been the inadequate assessment of external variability, suggesting that the commonly quoted errors do satisfactorily describe the internal precision but not the variation between laboratories.

Provisionally, we can attempt a breakdown of the components of variation across the three stages for each laboratory type. For both gas and liquid scintillation counting approximately 66% of the variation stems from Stage 1 (primarily counting process for LS labs, counting and synthesis for GC labs). For AMS labs, the major component of variation is introduced in Stage 3, which may be loosely ascribed to sampling and pretreatment. Given the nature of radiometric measurement and the small sample requirements of AMS labs, these general findings are not entirely unexpected.

Conclusions

In the first instance, it has been demonstrated that the internal precision of results is well covered by the quoted errors and that this is true for all laboratory types.

Secondly, evidence of systematic biases amongst the laboratories can be seen. The magnitude and sign of the biases have for some laboratories changed from stage to stage, indicating relatively short-term fluctuations in the source of variation in the results. In some instances, the causes of such fluctuations have been identified by the individual laboratory and corrected. We would emphasise that a main objective and benefit of the study was to identify and remedy errors in laboratory procedures. Thirdly, we find widespread evidence that the quoted errors do not adequately describe the variation amongst laboratories, even accounting for bias.

The observed degree of variation perhaps reflects difficulties in maintaining suitable and sufficient laboratory standards for calibration against primary laboratory standards such as oxalic acid. The type and level of pretreatment applied by individual laboratories vary considerably and for accelerator laboratories, this variability along with the question of representative sampling of the material, is clearly critical. The additional variation in results apparent for liquid scintillation laboratories may well reflect the increasing complexity of current technology.

Future Plans

In the light of the above findings, a detailed quality assurance proposal has been prepared by the ^{14}C community with the aims of:

a)   helping laboratories assure themselves that they are producing accurate data;

b)   alerting them when problems occur;
and
c)   ensuring the user's confidence in the results.

The proposals have two elements which when combined demonstrate the considerable commitment of the ^{14}C community to proving the high quality of its data.

Firstly, a recommended protocol for internal laboratory procedures for Quality Control is about to be published (in *Radiocarbon*). This will ensure a core set of monitored procedures, with corresponding documentation, in all laboratories. Although some differences will exist between laboratory procedures, an essential part of the protocol will involve the regular analysis of reference materials. Such a series of reference materials has for the first time been established, being held and distributed on request by the International Atomic Energy Agency in Vienna. These materials require no pretreatment by laboratories and, after the first inter-calibration, will be of known age. This stage of the proposals is already underway and samples are being distributed in May 1990, the results to be reported by December 1990 and a meeting and report to be held in February, 1991, in Vienna.

Secondly, in addition to the assays of such reference materials of known activity, regular blind inter-comparisons will be organised. These inter-comparisons will use natural sample materials, for which the ages will not be known in advance. They will thus be used to assess the effectiveness of internal laboratory procedures in improving and maintaining the quality of ^{14}C dates. The next inter-comparison is planned to begin in 1991, allowing the first round of standard samples to have been assayed. This phase of the international quality assessment programme will continue to be led from the UK at Glasgow University/NERC ^{14}C lab/SURRC, and will be funded jointly by SERC (who supported the previous study) and NERC.

These proposals represent an important step forward in assuring future user confidence in ^{14}C dates. They are a response to user pleas for demonstrably reliable ^{14}C dates. In addition, they provide tangible benefits to participant laboratories giving them absolute baselines on which to assess accuracy and allowing independent assessment of precision. In combination, the proposals herald a further development in the collaborative relationship between user and ^{14}C laboratory and demonstrate the continuing commitment of the ^{14}C community to ensuring the quality of its routine results.

References

COOK, G. T., HARKNESS, D. D., MILLER, B. F., SCOTT, E. M., BAXTER, M. S. & AITCHISON, T. C. (1990). International Collaborative Study-Structuring and Sample Preparation. *Radiocarbon*, 32 (3), pp 267-271.

SCOTT E M, LONG A & KRA R (1990). Proceedings of the Workshop on Intercomparison of Radiocarbon Laboratories. in press, *Radiocarbon*, 32 (3), 253-397.

SCOTT E M, BAXTER M S, AITCHISON T C, HARKNESS D D & COOK G T (1986). Announcement of a new collaborative study for intercalibration of ^{14}C dating laboratories. *Radiocarbon* 28, 167-169.

SCOTT E M, AITCHISON T C, HARKNESS D D, COOK G T, & BAXTER M S (1990). An overview of all three stages of the International Radiocarbon Intercomparison. in press, *Radiocarbon*, 32, 311-323

WATERBOLK H T (1983). Ten guidelines for the archaeological interpretation of radiocarbon dates. *Procs 14C and Archaeology*, Groningen, 1981. Pact 8, 57-71.

Quaternary Proceedings No. 1, 1991 5-10
© Quaternary Research Association, Cambridge

AMS Dating: Present Status and Potential Applications

R.E.M. Hedges

R.E.M. Hedges, 1991 AMS Dating: Present Status and Potential Applications, In *Radiocarbon Dating: Recent Applications and Future Potential* (ed. J.J. Lowe). Quaternary Proceedings No. 1, John Wiley & Sons Ltd, Chichester, pp. 5-10.

Abstract

The current performance of the Oxford AMS system, especially in relation to dating Quaternary sediments, is reviewed. This covers sample size, age range, accuracy and questions of cost, sample submission procedure, throughput and intercomparability.

An account of how AMS might best be applied to the 'dating' of Quaternary sediments is given. This requires detailed consideration of the different sources of carbon being dated, and outlines the possibilities for recognising and isolating a single source for measurement. Results from work at Oxford which aims to do this are used as illustrations of some of the problems.

KEYWORDS: dating; radiocarbon; sediment; AMS; Quaternary

Oxford Radiocarbon Accelerator Unit, Research Laboratory for Archaeology and the History of Art, 6 Keble Road, Oxford, OX1 3QJ, U.K.

Introduction

This paper considers AMS dating from two aspects; namely, the present operational 'state of play' of ^{14}C dating by AMS in the U.K., and what advantages the small sample requirement of AMS brings to the dating of environmental deposits.

Current operational aspects

Resources and administration

The Oxford Radiocarbon Accelerator Unit (ORAU) is the only AMS laboratory in the U.K. Until 1992 it will have been largely supported by SERC, through its Science-Based Archaeology Committee, and the predominant work of the Unit has been on archaeological issues. Environmental dating has been carried out under the following arrangements:-

(a) in relation to archaeology, and supported by the Programme Advisory Panel of the SERC Science-Based Archaeology Committee;

(b) as part of the Oxford laboratory's 'in-house' programme;

(c) as a commercial arrangement (£350 + VAT per date);

(d) as an application to the NERC Radiocarbon Laboratory Steering Committee;

(e) as part of a Research Grant Application to NERC, which would include the cost for dates by ORAU.

Technical capabilities of AMS dating at Oxford

The standard sample size dated is 1–3 milligrams of carbon.

Recently the development of the CO_2 source has meant that samples as small as 100 micrograms can be dated. However, such a small sample will give a lower precision (because fewer ^{14}C atoms can be detected) and any effect of laboratory contamination, which in general tends to be a more or less constant quantity, will be relatively larger, thereby making the date less accurate (especially for older dates). To some extent these limitations are inherent, but also they can be mitigated by improvements to laboratory technique, and work is continuing to this end (for example a special gas-handling system for very small samples is being designed). Another 'difficult' class of sample is where the carbon content is very low (<1%), since comparatively large quantities must be processed, and this is again liable to increase the incidence of laboratory contamination.

The error estimate (at one standard deviation) for fairly recent samples is 60 – 80 years, and this may increase to about 100 – 120 years at 12 ka. Beyond this period, the dominant error becomes the uncertainty in 'background', and this uncertainty at present limits the ultimate datable age to about 45 ka. In AMS measurements the 'background' is never due to an inability to identify a low count rate of ^{14}C, but comes from ^{14}C which may not have been originally in the sample. The uncertainty derives from several sources, but the main problem is the contamination of samples, ion source targets, or the ion beam by up to 0.5% 'modern' carbon. This error is relatively larger for smaller samples, and has an exponentially increasing effect with sample age.

Some comment about laboratory errors is in order, especially bearing in mind the result of the recent International Intercomparison (see Scott *et al.*, this volume). The techniques of AMS are more complicated than radiometric counting of ^{14}C, and for this reason the accuracy of AMS dating is unlikely to equal the 'high precision' results of the best of the radiometric laboratories. However, because most of the systematic errors are cancelled or reduced through making measurements on ısating isotopes, or by direct comparison with standards during the same run, systematic errors in AMS should be small (compared to the quoted error) and are of necessity frequently checked. Errors arise in the *measurement* because the ion beam from the sample is in some way sufficiently different from the ion beam from the standard that the mass spectrometric system acts to discriminate each isotope differently in each beam. The extent to which this happens in general appears to have a random component of about 0.25 – 0.5% per sample. Whether measurement errors follow a Gaussian curve (and in particular, whether the number of results in error at 3 sigma is thirty times less than at one sigma), is not so clear, and not so easily disclosed by the Intercomparison exercise. The distinction between random and systematic error is largely a reflection of how well a particular source of error is understood or controlled. Counting statistics, which give rise to an accurately estimated error, should ideally be the dominant error term, but generally a laboratory will try to mimimise its total error. Other error terms have to be estimated from experience of the system performance in circumstances where they can be quantified (for example, the measurement of known-age material). The 'Background', or laboratory contamination, correction is a case point. The correction to be made is systematic, but this correction is in fact only an estimate, and therefore contains its own, random error. (In the correction used at Oxford, the random error is estimated to amount to about half the correction.) One major potential source of error, not included here but which forms the basis of the discussion in the second part of this paper, is the contamination of the sample in its environment. By being more or less unique, such an error source is much more difficult to estimate from experience of past laboratory performance on known age material, and suitable warning as to possible errors in the date should be included in the publication.

Projects undertaken by the Oxford Unit

It may be of interest to summarise the nature and size of environmental dating projects carried out at Oxford up to 1989. Excluding faunal dating, a total of 152 dates were measured before 1987. Of these 90 were part of a D.Phil. study (Fowler, 1985) on the dating of sediments, and the results of that study will be referred to below. The remainder were distributed as shown in Table 1.

Table 1 Number of dates measured by Oxford Unit up until 1989

	Before 1987	Since 1987 excl. NERC dates	Since 1987 incl. NERC dates
Shell (marine)	7	17	17
Shell (terrestrial)	12		7
Sediments (lacustrine)	32	30	19
Sediments (peats, estuarine deposits)		26	
Sediments (marine)			6
Alluviation processes	11	32	

Present and future developments at Oxford

Apart from a continuing attempt to improve productivity, the main technical developments have centred on using the CO_2 source (Bronk & Hedges, 1989) and on improving combustion techniques. The CO_2 source has many advantages of convenience, and its ability to date very much smaller samples, when necessary, has already been noted. We hope for improvements in accuracy, although these are not so far apparent, and will require further work. It does mean that it is now possible to measure submitted samples of CO_2 from other laboratories, although we would warn interested laboratories that a major effort must go into validating the chemical processes by which CO_2 is produced before measurements on submitted gas can be regarded as a 'radiocarbon date'. As a cautionary example, nearly all projects with submitted samples that have had treatment in other chemistry laboratories have had super-modern contamination. This is in every case likely to be from previous use of ^{14}C tracer compounds in the laboratory, and demonstrates the susceptibility of the very small samples used in AMS to external contamination.

A second development is the method of oxidizing pre-treated material to CO_2. This is now accomplished by a commercial CHN analyser (Europa 'RoboPrep') in an automated way. The gas is purified by gas chromatography and an on-line $^{13}C/^{12}C$ measurement is also made, as well as measurements of the C and N content. This method has lower sample blanks (*i.e.* less modern C contamination), and is more reproducible, than our previous method of oxidizing with CuO. It is not readily adapted to samples containing less than 3% carbon without some degree of pre-enrichment of the carbon.

At present we have no significant research programme to investigate the potential for extracting different fractions, and to explore sophisticated pre-treatment methods in a systematic way. Such research is, however, essential if AMS dating is to realise its potential in environmental dating (as should become clear in the second part of this paper). This can perhaps best be achieved at this stage by collaborating both with workers with suitable field problems, and with chemists interested in understanding diagenetic processes in deposits containing organics. Otherwise, feed-back from different project experiences, coupled with application of the best available knowledge to pre-treatment methods, brings only slow progress.

In the second part of this paper, I hope to show how AMS can be useful, and where we are limited by our knowledge of the organic chemistry of what is being dated.

Advantages and use of AMS dating in environmental studies

Why date small samples?

AMS dating, per sample, is more expensive than 'conventional' or radiometric dating, and therefore needs additional justification. Also, as a technique, the measurement may not be as precise (although it is arguable that the 'date' produced is more accurate because a more appropriate sample has been measured).

Possible reasons are:-

(a) Large sample available, but at the expense of time resolution. (*i.e.* the sedimentation rate is comparatively low).

(b) The sample is extremely scarce (*e.g.* the collection of benthic foraminifera).

(c) The carbon content is very low. (Actually this may also pose problems for AMS); An example here is in the dating of atmospheric CO_2 trapped in polar ice cores.

(d) Where it is *preferable* to *select* a component from the sample. This might be some kind of macro/micro fossil, usually separable mechanically. Or it might be a chemical fraction, extracted by chemical procedures. Such selection overcomes many of the difficulties of dating deposits which have had a complex formation history.

The problems in radiocarbon dating

Working within the inevitable constraints of measurement accuracy and the effect (or lack) of the calibration curve, radiocarbon dating raises two fundamental problems. These are; whether the 'event' that is being dated can be associated with the environmental isolation of a set of carbon atoms; and to what extent the carbon atoms being measured correspond to the isolated set of carbon atoms associated with the 'event'. In many cases no problems arise. For example, the surviving protein in an animal bone can be chemically purified and freed from other carbonaceous contamination, its carbon atoms measured, and the result related to the death of the animal. Even here, the ability of AMS techniques to perform more stringent purification of bone protein has advantages in many cases. However, it is with the much more complex processes, particularly in the formation of terrestrial sedimentary deposits, that the interpretation of a radiocarbon 'date' on a 'sample' becomes highly problematic. This is illustrated in Figure 1 for the relatively complex system of lacustrine sedimentation. Sedimentation can be the result of a large number of possible sources of carbon, each of which is likely to have a different age. Furthermore, mixing processes of various kinds serve to confuse the sedimentary components before they are sampled. Many of the well-known difficulties in radiocarbon dating, the 'hard water effect', the finite age of recent sediments due to significant incorporation of terrestrial material from accelerated erosion, and the somewhat-too-old dates on earliest post-glacial sediments which have a high proportion of geologically dead graphite (Olsson 1972, Schoute *et al.* 1983) are apparent.

AMS ^{14}C dating has no single solution to these difficulties, but certainly can help in a number of ways. For example, the (dis)agreement between carefully chosen fractions can disclose the extent to which such problems are affecting dates in a particular situation. In general, however, undertaking radiocarbon dating by AMS underlines the need for fuller information on the sediment-forming process, both in terms of analyses of the sediment and in terms of the local palaeo-environment. This is necessary in order to assess the relative contributions from different organic sources, as well as subsequent diagenetic processes. Whether it is possible to isolate a single component from a sediment whose ^{14}C age can be guaranteed to give the date of formation of the deposit will almost certainly depend on the particular sediment, and in general is unlikely. The value of particular fractions that can presently be measured is discussed below.

Fractions from lacustrine sediments

Fractions studied so far include:-

— macrofossils
— lipids
— humics/fulvics
— humin : celulose : elemental carbon.

They are discussed in turn.

Macrofossils

Where these can be recovered (and scarcity is the main obstacle to their dating) they obviously constitute the sample of preference. Macrofossil remains frequently comprise less carbon than expected, and are fragile in their ability to withstand chemical pre-treatment, but otherwise offer no specific problems to AMS dating. The possibility that a macrofossil may be secondarily deposited must of course be considered, especially since samples robust enough to be selected are the ones more likely to survive redeposition.

Examples include insect remains (Hedges *et al.* 1989, OxA-1457) from the Condover mammoth, and Chironomid head capsules (OxA-2113) from Lough Neagh (V. Jones, to be published). The latter (requiring about 500 isolated individual capsules – although this requirement is now somewhat relaxed with the possibility to date smaller samples with the CO_2 source) offers promise to date sediments formed in the last millennium, where the contribution from soil erosion is serious. Birch fruits have been the subject of a major dating project on Swiss lakes (Zbinden *et al.* 1989). Pollen has also been dated (Brown *et al.* 1989) – albeit in rather special circumstances. Numerous dates have been obtained on material whose disintegration renders it less well characterised; this includes twigs, leaves, and seed remains. The advantages of macrofossils – that the material being dated is biologically and often specifically identifiable – suggest for the future a sampling strategy which explicitly aims to recover macrofossils for dating (*e.g.* at lake margins).

Lipids

These are defined as the class of chemical compounds which are soluble in common organic solvents, or mixtures of them, such as methanol/chloroform, and correspond to natural fats, waxes and their diagenetic products. Lipids tend to be more stable in the environment than other organic molecules (such as peptides and polysaccharides), and their insolubility in water makes them relatively immobile. Furthermore, many lipids are characteristic of the organisms which produced them, either as 'marker' compounds (for which a voluminous literature is accumulating – e.g. Cranwell 1982, Robinson *et al.* 1984 & Cranwell *et al.* 1987 and refs therein), or in rather general terms (for example, that long chain ($C_{25} - C_{33}$) alkanes, aldehydes, ketones and carboxylic acids derive from the waxes of higher plant cuticles).

The study of lipids in sediments divides into two. On the one hand is the highly detailed analysis of many hundreds of compounds, often present at the parts per billion level. This requires relatively sophisticated analytical equipment and the relevant expertise. On the other hand is the extraction of lipids into suitable 'classes' in quantities large enough for ^{14}C dating. These two activities should be co-ordinated so that the origins of the lipids used in dating can be ascertained as far as possible. In very general terms, analytical information should be able to estimate the higher and lower plant contributions and perhaps the degree of bacterial turnover.

When it comes to extracting lipids for radiocarbon dating it is only in very carbon rich sediments that it is feasible to isolate specific compounds. For example, Ellesmere provides

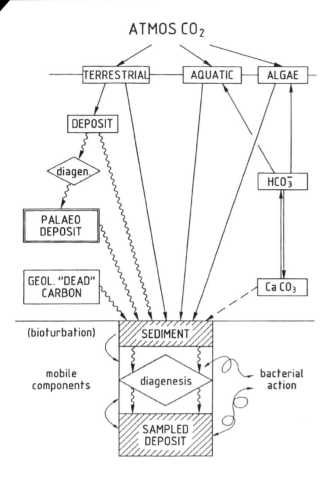

ATMOS CO$_2$

Figure 1 This diagram shows the possible routes by which atmospheric CO$_2$ is incorporated into the sediment column.

such a rich sediment (Farr *et al.* 1990), with the theoretical possibility to isolate individual compounds (*e.g.* nC$_{26}$ carboxylic acid and nC$_{27}$ alkane) known to come from higher plants. The carbon content corresponding to such material amounts to about 20 – 50 parts per million of the dry sediment, so that at least 25 g dry sediment is needed for a date. The total lipid component is, however, very much higher, and in many cases the most abundant components – usually the C$_{24}$-C$_{32}$ fatty acids – can be isolated and dated without too much difficulty. Therefore, although dating a lipid fraction is useful, because of its immobility and because it is chemically better characterised than most other components of a sediment, it may still have a complex origin and give an inappropriate date, and it will almost certainly require interpretation in the light of a detailed analysis of all the lipid compounds present.

It is worth stressing that so far relatively few lipid dates on sediments have been made, and that far "more work needs to be done" both in terms of what is practical to date and in terms of how to interpret lipid analyses.

Humics and Fulvics

These materials are defined as being extractable in dilute alkali. An extensive literature exists (Schnitzer 1978) but it is clear that they are not at all well characterised chemically, that they can originate from almost any thermodynamically less stable organic species, and that they are potentially highly mobile in sediments. They also contribute a high proportion of the carbon (20. – 50 %) in many sediments, and so are

inevitable candidates for ^{14}C dating. Our own experience, from dating carbonaceous material in mainly archaeological deposits (Batten *et al.* 1986) is that the humic component frequently does reflect the 'age' of the deposit. However, examples abound where humic dates appear to be too young, and the potential mobility of humics is held to blame (Head *et al.* 1989). Clearly much depends on the site, on its hydrological regime, and on the sediment mineralogy (for example, humics may be strongly bound to clay particles and thereby immobilised).

Insoluble carbon

Humin is generally defined as insoluble but humic-like material, which may grade chemically into graphite. It may be oxidised by treatment with sodium hypochlorite, which is the standard method to remove lignins from cellulose. Although in theory it should be possible to extract and hence specifically purify cellulose, this has not been done, because of the elaboration of the chemistry involved and also because partially degraded cellulose does not behave in a predictable way. However, a positively extracted and identified cellulose fraction would be a useful addition to the dateable components from sediments. The same could be true of lignins, which are likely to be present in material derived from terrestrial erosion. No work has been done on this; the chemical isolation of lignins from sediments would be difficult. A first step could come from the analytical study of sediments, where the lignin contribution could be quantified. Again, this has not been carried out. Present methods are limited to extracting a 'total resistant humin + cellulose + graphite' component (using hydrofluoric acid), and a 'cellulose + graphite' fraction (by treating the HF extract with NaOCl). This can be revealing, and an extreme case is given in the immediate post-glacial sediments of Llyn Gwernan (Lowe *et al.* 1988) where the carbon content is very low and seems to consist largely of geological graphite. This is very evident in the much older ^{14}C age of the 'humin' fraction in comparison with the other fractions.

A related problem, that of measuring the ^{14}C in finely disseminated graphite in sediments, has been tackled through the use of different oxidising agents (including nitric acid with potassium chlorate) (Gillespie 1990).

A summary

The nature of the information that different fractions offer can be summerised as follows:-

1 Any macrofossils should be dated and can usually provide sufficient dating evidence by themselves.

2 Other crudely characterised fractions (e.g. 'lipid'; 'humics'; 'insoluble') can be measured for ^{14}C to determine their consistency. If different ages are produced, the possible causes may be further pinned down and different C sources quantified by measuring a chemically better defined fraction, but this may not be possible in most cases, since many sediments do not contain sufficient suitable material. On the other hand, detailed qualitative analysis of the sediment lipids can provide information which should help to interpret both the ^{14}C age of various fractions and the formation and diagenesis of the sediments.

3 Although chronological information on sediments is always likely to be incomplete, the potential information from the isotopic composition is considerable. However, very little

comprehensive research has been carried out, partly because of the cost in dating many fractions, because of the recalcitrant chemistry of much of the organic material, and partly because each deposit presents a rather different situation. A suggested 'algorithm' for dating sediments was published by Fowler *et al.* in 1986, and the situation has changed very little since then.

Comments on various processes in lacustrine sediment formation

Hard water effect

In regions where the country rock contributes significantly to the carbonate ions in groundwater, aquatic plants (including algae) will incorporate a fraction of 'dead' carbon during photosynthesis. This fraction is variable, although in extreme cases could be up to 50%. (This assumes complete equilibrium between rock carbonate and dissolved atmospheric CO_2, and no direct atmospheric uptake by the plants.) Once photosynthesized, the 'dead' carbon contribution is very difficult to quantify; for example, the stable isotope composition will not necessarily differ from that of carbon obtained from the atmosphere. The only hope of dating samples from sediments formed in 'hard water' appears to be to use terrestrial macrofossils. A detailed understanding of the 'hard water' status of the water column at the time of deposition (which can be partly inferred from the sediment itself) can act as a guide. The sediment may contain biogenic carbonate, and if this has the same ^{14}C date as, say, the lipid component, it is unlikely that a hard water effect is operating.

The terrigenous contribution

It is quite feasible to isolate a terrigenous lipid contribution using organic chemical methods alone, but the distinguishing of reworked and non re-worked components can be problematical (although other sedimentological studies – *e.g.* magnetic susceptibility – may well indicate changes in the terrigenous contribution). As mentioned above, the presence of lignins is also likely to indicate terrigenous material, while lipid markers characteristic of algae (algal sterols such as dinosterol and gargosterol) represent a carbon input that is clearly autochthonous. (But it is not possible to isolate an 'algal' fraction in sufficient quantity from a depth (time) interval of a typical sediment core small enough to be useful for AMS-^{14}C dating.) As an example, see the discussion on the radiocarbon dates from Lake Zacapu, Mexico in Fowler *et al.* 1985.

Diagenesis and bacterial action

The general effect of diagenesis is to reduce the amount of information conveyed by the chemical species composing the sediment, for example, preventing the extraction of well characterised compounds. The formation of humic acids, which to some extent must migrate upwards as the sediment is compressed, is a potential complication (although so far this has not been identified in the field). Continuing diagenesis in peats, including the formation of CO_2 through fermentation, is likely to be a problem in addition to those already recognised. (Peats are also more susceptible to 'mixing' from rootlet growth.)

Most bacterial action within the sediment also has the effect of contributing to the general loss of information on the specific inputs, and has little effect on carbon dating because the carbon atoms are re-cycled. However, this re-cycling could involve transport of metabolites (such as methane) up the sediment column, and so is equivalent to a form of molecular bioturbation. Bacterial action is often revealed in the production of branched-chain fatty acids. Extensive bacterial action can be identified in this way, but so far its effects on ^{14}C dating are only surmised. What is quite clear is that cores stored in the atmosphere (at least at room temperatures) may 'fix' substantial quantities of atmospheric CO_2 by bacterial action, and this has the potential for serious errors in dating.

Conclusions

Radiocarbon measurement by AMS is now applicable to the full range of sedimentary deposits encountered in environmental dating. The fundamental reason for choosing AMS techniques (over beta counting) is the thousandfold or more reduction in sample size. The selectivity made possible with a smaller sample size is particularly valuable in trying to date sediments formed from a complex of processes which contribute organic material of different radiocarbon ages. Whether AMS dating can give a definite 'date' in a particular case depends on the availability of particular components, and requires some understanding of the organic geochemistry of the sediment. However, in almost all circumstances AMS dating of different fractions enables the reliability of a given radiocarbon 'date' to be much more accurately assessed.

References

BATTEN, R. J., GILLESPIE, R., GOWLETT, J. A. J. & HEDGES, R. E. M. (1986). The AMS dating of separate fractions in archaeology, *In* Stuiver, M. & Kra, R. S. (eds) 12th International Radiocarbon Conference, Trondheim, Norway 1985, Proceedings, *Radiocarbon* 28, 698-701.

BRONK, C. R. & HEDGES, R. E. M. (1989). Use of the CO_2 source in radiocarbon dating by AMS. *Radiocarbon* 31, 970-975.

BROWN, T. A., NELSON, D. E., MATHEWES, R. W., VOGEL, J. S. & SOUTHON, J. R. (1989). Radiocarbon dating of pollen by accelerator mass spectrometry. *Quaternary Research* 32, 205-212.

CRANWELL, P. A. (1982). Lipids of aquatic sediments and sedimenting particulates. *Progress in Lipid Research* 21, 271-308.

CRANWELL, P. A., EGLINTON, G. & ROBINSON, N. (1987). Lipids of aquatic organisms as potential contributors to lacustrine sediments. II. *Organic Geochemistry* 11, 513-527.

FARR, K. M., JONES, D. M., O'SULLIVAN, P. E., EGLINTON, G., TARLING, D. H. AND HEDGES, R. E. M. (1990). Paleolimnological studies of laminated sediments from Shropshire-Cheshire meres. *Hydrobiologia* (in press).

FOWLER, A. J. (1985). *Radiocarbon dating of lake sediments and peats by accelerator mass spectrometry*, D.Phil thesis, Oxford.

FOWLER, A. J., GILLESPIE, R. & HEDGES, R. E. M., (1986).

Radiocarbon dating of sediments, *In* Stuiver, M. & Kra, R. S. (eds) 12th International Radiocarbon Conference, Trondheim, Norway, 1985, Proceedings, *Radiocarbon* 28, 441-450.

GILLESPIE, R. (In Press). On the use of oxidation for AMS sample pretreatment, *In* Yiou, F. & Raisbeck, G. (eds) AMS 5: Fifth International Conference on accelerator mass spectrometry, Paris 1990. *Nuclear Instruments and Methods*, in press.

HEAD, M. J., ZHOU, W. & ZHOU, M. (1989). Evaluation of ^{14}C ages of organic fractions of paleosols from loess-paleosol sequences near Xian, China. *In* Long, A., Kra, R. S. & Srdoc, D. (eds) 13th International Radiocarbon Conference, Proceedings, *Radiocarbon,* 31, 680-694.

HEDGES, R. E. M., HOUSLEY, R. A., LAW, I. A. & BRONK, C. R. (1989). Radiocarbon dates from the Oxford AMS system: Archaeometry datelist 9, *Archaeometry* 31, 207-234.

LOWE, J. J., LOWE, S., FOWLER, A. J., HEDGES, R. E. M. & AUSTIN, T. J. F. (1988). Comparison of accelerator and radiometric radiocarbon measurements obtained from Late Devensian Lateglacial lake sediments from Llyn Gwernan, North Wales, UK., *Boreas* 17, 355-369.

OLSSON, I. U. (1972). A critical analysis of ^{14}C dating of deposits containing little carbon. *In* T. A. Rafter & T. Grant-Taylor, (eds.) International ^{14}C Conference, 8th, Proceedings: Wellington, *Royal Soc. New Zealand*, P G11-G18.

ROBINSON, N., CRANWELL, P. A., FINLAY, B. J. & EGLINTON, G. (1984). Lipids of aquatic organisms as potential contributors to lacustrine sediments. *Organic Geochemistry* 6, 143-152.

SCHNITZER, M. (1978). Humic substances: chemistry and reactions. *In* (M. Schnitzer & S. U. Khan, Eds.) *Soil Organic Matter*, Elsevier, New York, pp. 1-58.

SCHOUTE, J. F. TH., MOOK, W. B. & STREUERMAN, H. J. (1979). Radiocarbon dating of vegetation horizons. *In* W. G. Mook & H. T. Waterbolk, (eds.) Groningen Conference on ^{14}C and archaeology, Proceedings: *PACT*, p295-311.

ZBINDEN, H., ANDREE, M., OESCHGER, H., AMMANN, B., LOTTER, A., BONANI, G. & WÖLFLI, W. (1989). Atmospheric radiocarbon at the end of the last glacial: An estimate based on AMS radiocarbon dates on terrestrial macrofossils from lake sediments. *In* Long, A., & Kra, R. S., (eds). 13th International ^{14}C Conference, Proceedings, *Radiocarbon* 31, 795-804.

Quaternary Proceedings No. 1, 1991 11-17
© Quaternary Research Association, Cambridge

First-Order ^{14}C Dating Mark II

C Vita-Finzi

C Vita-Finzi, 1991 First-Order ^{14}C Dating Mark II, in *Radiocarbon Dating: Recent Applications and Future Potential* (ed. J.J. Lowe). Quaternary Proceedings No. 1, John
Wiley & Sons Ltd, Chichester, pp. 11-17.

Abstract

First-order ^{14}C dating was developed to facilitate quantitative neotectonic analysis on coasts. The technique relies on liquid scintillation counting of absorbed CO_2 derived from carefully pretreated shell carbonate. It usefully complements conventional ^{14}C assay in the construction of Holocene deformation chronologies and their use in structural interpretation.

KEYWORDS: ^{14}C; first-order; dating, shell; neotectonics

Department of Geological Sciences, University College London, Gower Street, London WC1E 6BT

First-Order Assay

The high cost of radiocarbon dating and the delays to which it is often subject have long hampered those aspects of Late Quaternary research that hinge on numerical ages and time correlation. Neotectonics is one such field of study, as it needs to quantify the progress of deformation in relation to other environmental changes and the information is especially welcome if it can be obtained during the investigation rather than months or years after it is over. The technique discussed here was developed following a series of field studies which could not be investigated with optimum briskness and thoroughness because the small number of ^{14}C ages that could be allocated to the work sometimes took over 2 years to be processed. As its title indicates, this paper reviews changes in procedure introduced since the method was first described (Vita-Finzi 1983).

Early in the search for an economical and fast alternative mode of ^{14}C assay that could be carried out in a normal earth science laboratory it became clear that procedures based on gas proportional counting or on liquid scintillation counting using benzene did not offer much scope for economy: sample preparation could not be simplified without jeopardising quality, and counting could not be cut short without increasing the margin of error above the currently favoured threshold of 10%.

Previous attempts to speed things up had not had much success. Libby (1970) himself had proposed the use of large counters to cut down on the time required to accumulate 10,000 or so disintegrations and thus increase the number of dates produced by ^{14}C laboratories, and Noakes *et al.* (1970) showed how a portable laboratory could be used in the field, but the suggestions did not arouse much enthusiasm and in any case they did not reduce the time or equipment required to prepare the samples.

Discussion with biochemists pointed the way to a novel approach. In many of their studies ^{14}C is used as a tracer and its assay is therefore a routine matter. Various simple techniques had been in use for several years for trapping CO_2, whether expired or obtained by oxidising tissue by dry or wet methods, in a liquid such as hyamine, toluene, ethanolamine or phenethylamine (Passmann *et al.* 1956; Horrocks 1968; Jeffay & Alvarez 1961; Schramm & Lombaert 1963). The apparatus ranges from a counting vial with a simple moisture trap to fully automated rigs capable of processing several samples a day (Horrocks 1974).

Sample preparation

Eichinger *et al.* (1980) then applied the method to hydrogeological samples and, in order to raise the carbon content of the sample, they constructed a special measuring chamber cpabale of counting volumes of up to 160 ml (that is 8 times the usual), equivalent to 5.3 g of carbon. Their experimental detection limit was 1% modern, corresponding to 28,000 years. In their rig the CO_2, which had been stored in a cylinder, was bubbled through a 1:1 mixture of absorbing solution and scintillator, with absorptions of over 95% being recorded. Each sample cost about $3.00 in consumables at 1980 prices.

The system was adapted for neotectonic research as follows. First, as emphasis was to be placed on carbonates, combustion could be avoided by treatment with HCl (initially 10% and later 50%) which in turn ensured vigorous CO_2 generation, active bubbling and hence good absorption without the need for a vacuum line or storage of the gas at low temperature. Second, as standard counters had to be used reliance was placed on standard 20ml vials. The absorption liquid remained a 1:1 mixture of Permafluor V and Carbosorb.

The first version of the rig followed the original in enclosing the absorption vessel in a water jacket and drying the CO_2 with water traps cooled with acetone and dry ice. Sample preparation took up to 2 hours (Vita-Finzi 1983). In due course the water jacket was abandoned, as the cooling merely slowed down the reaction. Total absorption is now accomplished in 30-40 minutes. Similarly, flushing of the mixture with nitrogen was found to be unnecessary and possibly detrimental. The system is now flushed out by generating CO_2 for a few seconds before connecting up the absorption vessel. To avoid exchange with the atmosphere when dispensing the saturated mixture the reaction is carried out in the counting vial with the help of a stainless steel cap (designed by S.J. Phethean) which screws onto the vial (Fig.1). In order to minimise background activity low potassium glass vials are used.

As vigorous bubbling may lead to splashing and hence the loss of some of the mixture through the outlet the favoured procedure is to absorb the CO_2 in a mixture composed of 10 ml

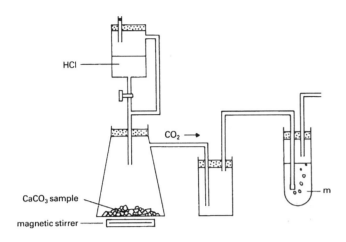

Figure 1 Simplified rig used to prepare samples for liquid scintillation counting. The contents of the trap are discussed in the text.

of Carbosorb and 5 ml of Permafluor and to add the missing 5 ml of Permafluor once the reaction is complete. (Bubbling into pure Carbosorb is unworkable as saturation renders the Carbosorb extremely viscous.) Where 10 ml samples are to be made there is of course no need for additional Permafluor after saturation.

Saturation can be recognised in three ways. First, the temperature of the mixture begins to fall, signalling the end of the reaction. Second, reweighing of the vial and its contents reveals a gain in weight of about 1.26 g per 10 ml of mixture, which corresponds with the 1.3 g stated by the manufacturers to be the maximum absorption level. Third, the channel ratios on the counter correspond to those obtained experimentally by bubbling CO_2 through a standard mixture until there is no further weight gain.

The first samples were run using 10% HCl; 50% is now favoured to minimize reaction time. Again, the original water traps have been scrapped for they sometimes became blocked, risking the loss of the sample, and the risk was in no way matched by the effects of small amounts of water on count rates. Hydrochloric acid vapour has a more serious effect not least by its effect on pH and a trap containing a solution of silver nitrate proved an effective remedy but a simpler alternative which also dries the gas is a tube packed with the desiccant magnesium perchlorate. So far as can be judged, simplification of the rig has not had any deleterious effect on sample quality.

The first counts were made on a Packard Tri-Carb 3255 using preset ^{14}C windows and 10 ml samples. As the machine was being used by chemists and biochemists for very active samples its background was high (17.7 cpm). Even so the results were encouraging: samples of known age spanning the last 7000 years gave results which were consistent with the ^{14}C half-life decay graph at 2 s.d. Counting was continued for 1000

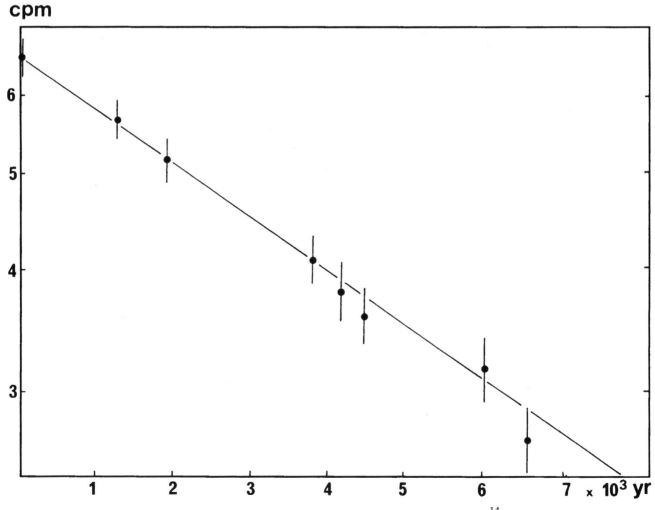

Figure 2   Comparison of conventional and first-order ages (Table 1). The line shows sample activity in accordance with the ^{14}C half-life and with the modern standard as origin.

minutes and counts per minute (cpm) were corrected for efficiency by using the External Standards Channels Ratio (ESCR) method.

The results were sufficiently encouraging (Vita-Finzi 1983) to justify further experiments. Some of the counting was now done on a dedicated but aged counter, a Nuclear Enterprises 8312, but heat generated by the counter was found to produce erratic count rates especially if the samples displayed any colour quenching, and reliance was once again placed on 'borrowed counting time' on various machines until access was gained to two machines designed by Packard with low level counting in mind, a single-sample Tri-carb 1050 and a multi-sample, refrigerated Tri-carb 2260XL. And although the highest counting efficiencies were reported to be obtained with 12ml samples, our own experiments in the end favoured 20ml samples because that volume ensured the largest range of cpm values between modern and background standards and, by almost filling the vial, reduced the problems created by photomultiplier cross-talk.

A set of activity measurements obtained on the 1050 is plotted in Figure 2 against ages obtained on the same samples by conventional methods. The activities are expressed in counts per minute (cpm) after subtraction of a background value of 7.8 cpm (B in Table 1). The results agree within 1 s.d. for the last 7000 years. The lower background activity and greater stability of the 2260XL, which is refrigerated, should extend the range of first-order dating towards the 21,000 year limit attained by Qureshi *et al.* (1989) with a refined version of the absorption method. Note that in Fig. 2 the ages are plotted in radiocarbon years. Tree-ring calibration would need to be applied to these values if calendric ages were at issue; similarly, the apparent age of modern marine molluscs, which usually amounts to a few centuries but locally exceeds 2500 years, is another factor which needs to be allowed for specially if interregional comparisons are being made.

Table 1 Comparison of first-order cpm values with conventional ¹⁴C ages

sample	material	cpm*	s.d.	conventional age (¹⁴C yr BP)	lab. no.
M	*Conus sp.*	14.6	0.12	c.AD 1850	
B (G82/4)	*Pecten jacobaeus*	7.8	0.08	> 42,000	SRR-2461
T1	*Estonia cf rugosa*	11.8	0.11	4200+90	HAR-1754
Wylfa	*tufa*	13.2	0.11	1890+70	Beta-18211
S81/6	*Pinctada margaritifera*	11.6	0.11	4460+60	Beta-2682
ETIII e	*Estonia rugosa*	10.7	0.10	6535+160	SRR-1312
J85/2	*Tridacna sp.*	12.1	0.11	3775+80	Beta-14773
SHE	*Circe arabica*	11.2	0.11	6020+80	Beta-2533
JTG	*Cardium sp.*	13.7	0.12	1265+100	Birm-244b

* measured on Packard 1050 for 1000 mins (1.0-20 keV channel), uncorrected cpm. B(G82/4) provides the background value subtracted from the other determinations to obtain the cpm used in Fig.2

The cost of consumables (chemicals and counting vial) per sample at 1990 prices is £0.4. One could argue that this igure is misleading because it disregards labour, overheads,

depreciation of the counter and other important items. But as the work can be done on borrowed equipment by teaching staff in addition to their existing duties the price is realistic.

Sample selection and pretreatment remains the key to dependable ages and justifies the belief that the results are more accurate if less precise than many of those produced by conventional means on poorly screened material. Samples are inspected by light and scanning electron microscopy as well as XRD analysis for evidence of contamination (Figs.3, 4 and 5). If any overgrowths or recrystallised zones are near the surface they are removed by mechanical abrasion followed by acid leaching. When possible sampling is confined to aragonitic samples as secondary calcite is readily detected by XRD. All the above screening techniques bear on small subsamples of the material to be dated and it is therefore important to select representative portions of the shell. Where any doubts persist, stable isotope analysis of an aliquot of the gas prepared for dating can be used to obtain a measure of the quality of the entire sample. At present this stratagem is not consistent with first-order dating because the CO_2 is not stored before being absorbed, but laboratories equipped with mass spectrometers could analyse the gas from sub-samples from different parts of the shell or indeed abstract some of the CO_2 as it is being generated.

Deformation Chronologies

The above method, like conventional ¹⁴C dating (Chappell & Polach 1972; Mangerud 1972; Pirazzoli *et al.* 1985), may be used to date the carbonate of molluscs, barnacles and corals whose relationship to the contemporaneous shore can be established and hence to evaluate displacements or deformation of the waterline. The ecology and depositional history of the dated material evidently colour interpretation of the dates. When a species has a wide depth tolerance it may be a better helpful indicator redeposited than in growth position (Richards 1982); again, the age of abrasion platforms are sometimes more informative than that of the coral into which they have been cut.

Speed and cheapness here come into their own. First, any attempt to identify regional deformation in detail requires sufficient data points for both general trends and departures from such trends to emerge. Second, where the date of the shoreline deposit in question is on redeposited fossils several age determinations are required to identify the main contributory sub-populations. Third, a surprising result may invite additional collecting at existing sampling sites or a switch to novel locations.

Nevertheless regular testing of the procedure with samples of known age, such as those illustrated in Fig. 2, is advisable. The need to date material other than shell (mangrove wood, for example: see Table 2) may provide the opportunity to do so constructively. The argument may be illustrated with examples from Indonesia, where an attempt is being made to link Holocene deformation to plate kinematics on the one hand and seismicity on the other.

The first study area comprised the islands of Simeulue and Nias, which form part of the Sumatran outer arc (Fig. 6). Previous workers had reported raised reefs on both islands. The sole radiometric age was a Th/U determination of 6500 ± 500 years BP on coral from a 3m terrace on Nias (Karig *et al.* 1980).

The first application of the method was to date the lowest reef or fossil intertidal platform at different locations on the two islands in order to discover when the last phase of emergence had taken place and whether it had affected all areas uniformly.

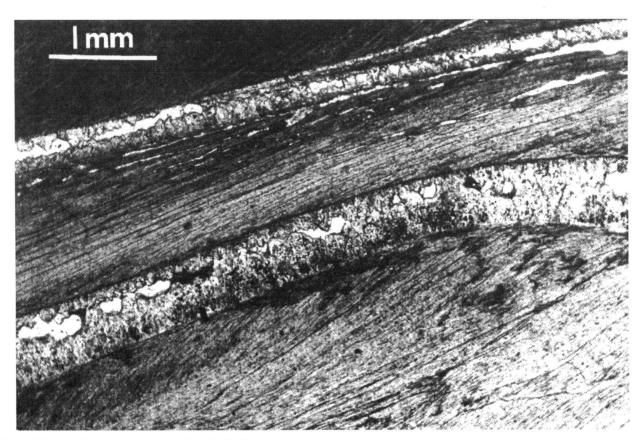

Figure 3 Foliated calcitic structure in *Ostrea* with void infilled by secondary calcite crystals.

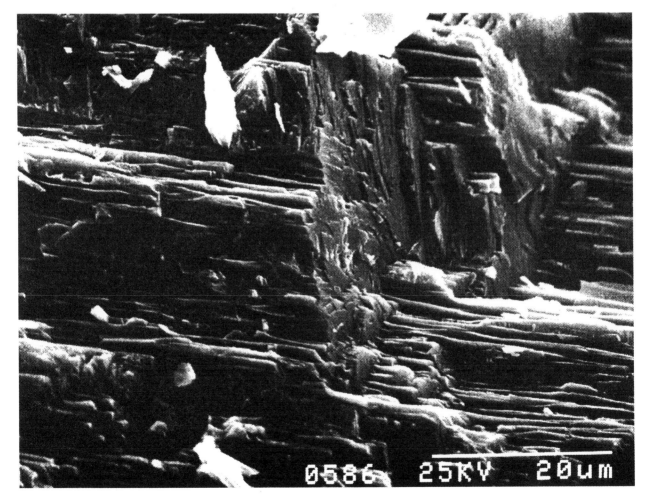

Figure 4 Scanning electron micrograph of *Astraea rugosa* showing unaltered aragonitic structure

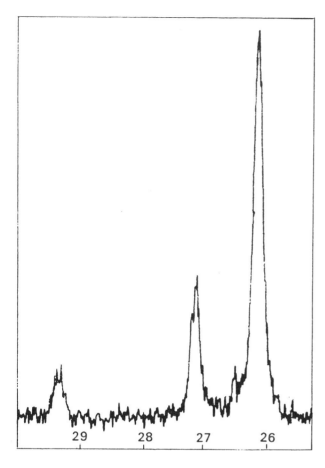

Figure 5 X-ray deffractogram of aragonitic shell contaminated with
secondary calcite (small peak on left)

inclined to carry out detailed stratigraphic and palaeontological work before sacrificing precious funds to radiometric dating. In some places, of which the Chilean coast is a fine example, without radiometric help it is difficult to distinguish between Pleistocene and Holocene deposits where they contain identical, well preserved shell faunas. In addition, sections displaying more than one well defined abandoned waterline were sampled in some detail to see how far movement was unidirectional and to identify its rate of progress.

Nineteen determinations were made on shell from emergent beach deposits and platforms on the two islands. At some sections there were several successive terraces above the modern intertidal platform. Table 2 shows that ages from two of the sections on the western coast of Simeulue are consistent with progressive emergence at about 0.2 – 0.5 mm/yr. As there is morphological and historical evidence for submergence on the east coast of the island the first-order dates could be held to indicate northeastward tilting. The data also indicate that at one of the sections on Nias, north of Gunongsitoli, subsidence at about 0.7mm/yr was followed by 1-2m of uplift. The sequence of events parallels those that preceded the 1923 earthquake at Kanto, south of Tokyo (Scholz & Kato 1978).

Taken as a whole the first-order ages for the two islands fall into three groups which, on the basis of tree-ring calibration (Klein *et al.* 1982), span 0 – 900, 2100 – 3800 and 5200 – 6600 yr BP. Uplift has apparently been a predominantly seismic process manifested during active periods separated by phases of quiescence of similar duration (Vita-Finzi & Situmorang 1989). The ages in Table 2 suggest that movement took place in increments of 0.5 – 0.8m, a view that gains support from the presence of up to 5 successive and progressively more dissected fossil intertidal platforms.

A plausible mechanism is the reverse faulting of trench deposits associated with shortening of the forearc stemming

Table 2 Selected first-order ¹⁴C dates from Nias and Simeulue

Site	height (m)	age (¹⁴C yr BP)	calib*	lab no
SIMEULUE				
Lakubang	1.0	2450±250	2590	UCL-145
	1.5	4500±400	5160	UCL-142
Ulung Kahat	1.7	3100±300	3360	UCL-95
	2.5	5400±350	6260	UCL-92
	2.5	5400±400	6260	UCL-91
NIAS				
Gunongsitoli	1.5	5800±300	6610	UCL-86
		>12,000		UCL-152
	0.5	4900±300	5680	UCL-87
		6600±300	7520	UCL-153
		5580±110	6455	Beta-23930 #

* Tree-ring calibration after Klein *et al.* 1982
# on mangrove wood

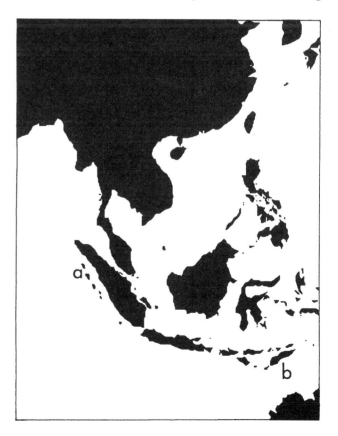

Figure 6 Location of Nias and Simeulue (a) and Timor (b) (Indonesia).

Trivial though these questions appear they are commonly answered too late in the enquiry because the investigator is

from subduction. The record of the last 300 years (Newcomb & McCann 1987) shows that a large proportion of interplate slip in the western part of the Sunda Arc occurs seismically. A great earthquake in 1861 (M_w = 8.3 – 8.5) was accompanied by uplift on the N and W coasts of Nias, where reefs and rock became exposed, just as there was submergence in southern Simeulue during an earthquake in 1907 (Hennessey 1971, 280).

The island of Timor (b on Fig. 6) also displays uplifted reefs. In the western part of the island, Jouannic et al. (1988) obtained a Th/U age of 152,000 ± 10,000 years from a coral head in growth position in the upper part of the fifth step at 44m in a flight of 7 terraces at Cape Namosain, and derived from it an uplift rate of 0.3mm/yr. This slow rate was supported, in their view, by numerous large modern reef platforms and by very limited mid-Holocene emergence. They also reported a Th/U date of 124,000 yr on the lowest (7m) of four emerged terraces in the nearby island of Semau.

As the Indo-Australian and Eurasian Plates are converging at about 7 – 8 cm/yr these results are puzzling. The puzzlement increases when we consider the micropalaeontological evidence for early Pliocene uplift at about 3mm/yr and at about half that rate for later periods (Audley-Charles 1986). In order to ascertain whether uplift had indeed decreased even more drastically by the Holocene, samples were collected for first-order dating from the lowest marine terraces in the areas already investigated by Jouannic et al.(1989).

All the first-order ages determined thus far on rock-cut terraces less than 3m above High Water are beyond the 12,000-year limit of the existing equipment. The two exceptions are on Tridacna and Anadara sp. from a beach sand 1m above High Water 5 km east of Kupang, which gave first-order ages of 4750 ± 250 (UCL-154) and 4400 ± 250 (UCL-155) respectively. Granted that the sand represents a genuine relative stillstand (and the absence of a corresponding wave-cut platform suggests it was shortlived) the emergence rate it indicates, namely 0.2 mm/yr, tallies with the Th/U results of Jouannic et al.(1988).

The next step will be to discover whether other parts of the island, including its interior, display a similar trend and if so when it set in. The study will thus need conventional ^{14}C and U-series ages in order to encompass a sufficiently long record. But it will also need additional first-order dates so that the survey can be extended around Timor and to its neighbours.

Acknowledgements.

I am indebted to Henry Polach, Doug Harkness, Richard Burleigh and Roger Marchbanks for advice, Fred Pearce and Derek Banthorpe for access to Liquid Scintillation counters, Stuart Phethean (inventor of the Phethean cap), Deborah Lewis, Tony Osborn and Tom Blyth for assistance and Ron Cooke for support.

References

AUDLEY-CHARLES, M.G. (1986). Rates of Neogene and Quaternary tectonic movements in the southern Banda Arc based on micropalaeontology. Journal of the Geological Society of London, 143, 161-175.

CHAPPELL, J. & POLACH, H.A. (1972). Some effects of partial recrystallisation on 14C dating Late Pleistocene corals and molluscs. Quaternary Research, 2, 244-252.

CHAPPELL, J. & VEEH, H.H. (1978). Late Quaternary tectonic movements and sea-level changes at Timor and Atauro island. Bulletin of the Geological Society of America, 89, 356-368.

EICHINGER, L., RAUERT, W., SALVAMOSER, J. & WOLF, M. (1980). Large-volume liquid scintillation counting of carbon-14. Radiocarbon, 22, 417-427.

HENNESSEY, S.J. (1971). Malacca Strait and West Coast of Sumatra Pilot (5th ed.). Hydrographer of the Navy, Taunton.

HORROCKS, D.L. (1968). Direct measurement of $^{14}CO_2$ in a liquid scintillation counter. International Journal of Applied Radiation and Isotopes, 19, 859-864.

HORROCKS, D. L. (1974). Applications of Liquid Scintillation Counting. Academic Press, New York.

JEFFAY, H. & ALVAREZ, J. (1961). Liquid scintillation counting of carbon-14. Analytical Chemistry, 33, 612-616.

JOUANNIC, C., HOANG, C.-T., HANTORO, W.S. & DELINOM, R.M. (1988). Uplift rate of coral reef terraces in the areas of Kupang, West Timor: preliminary results. Palaeogeography, Palaeoclimatology, Palaeoecology, 68, 259-272.

KARIG, D.E., LAWRENCE, M.B., MOORE, G.F. & CURRAY, J.R. (1980). Structural framework of the fore-arc basin, NW Sumatra. Journal of the Geological Society of London, 137, 77-91.

KLEIN. J., LERMAN, J.C., DAMON, P.E. & RALPH, E.K. (1982). Calibration of radiocarbon dates. Radiocarbon, 24, 103-150.

LIBBY, W.F. (1970). Ruminations on radiocarbon dating. IN: Olsson, I.U. (ed.), Radiocarbon Variations and Absolute Chronology, 629-640. Wiley, New York.

MANGERUD, J. (1972). Radiocarbon dating of marine shells, including a discussion of apparent age of Recent shells from Norway. Boreas, 1, 143-172.

NEWCOMB, K.R. & McCANN, W.R. (1987). Seismic history and seismotectonics of the Sunda Arc. Journal of Geophysical Research, 92, 421-439.

NOAKES J.E., SCHNEIDER, K.A. & BRANDAU, B.L. (1972). A mobile archaeological laboratory: microsample extraction and radiocarbon dating. Proceedings of the 8th International Radiocarbon Conference, Wellington, 203-208.

PASSMANN, J.M., RADIN, N.S. & COOPER, J.A.D. (1956). Liquid scintillation technique for measuring Carbon-14 dioxide activity. Analytical Chemistry, 28, 484-486.

PETRIE, W. M. FLINDERS. no date. Ten Years' Digging in Egypt (1881-1891). The Religious Tract Society, London.

PIRAZZOLI, P.A., DELIBRIAS, G., KAWANA, T. & YAMAGUCHI, T. (1985). The use of barnacles to measure and date relative sea-level changes in the Ryukyu Islands, Japan. Palaeogeography, Palaeoclimatology, Palaeoecology, 49, 161-174.

QURESHI, R.M., ARAVENA, P., FRITZ, P. & DRIMMIE, R. (1989). The CO_2 absorption method as an alternative to benzene synthesis method for ¹⁴C dating. *Applied Geochemistry*, 4, 625-633.

RICHARDS, G.W. (1982). Mediterranean Intertidal Molluscs as Sea-level Indicators. PhD thesis, London University.

SCHOLZ, C.H. & KATO, T. (1978). The behaviour of a convergent boundary: crustal derormation in the South Kanto District, Japan. *Journal of Geophysical Research*, 83, 783-797.

SCHRAMM, E. & LOMBAERT, R. (1963). *Organic Scintillation Detectors*. Elsevier, Amsterdam.

VITA-FINZI, C. (1983). First-order ¹⁴C dating of Holocene molluscs. *Earth and Planetary Science Letters*, 65, 389-392.

VITA-FINZI, C. (1987). ¹⁴C deformation chronologies in coastal Iran, Greece and Hordan. *Journal of the Geological Society of London*, 144, 553-560.

VITA-FINZI, C. & SITUMORANG, B.(1989). Holocene coastal deformation in Simeulue and Nias, Indonesia. *Marine Geology*, 89, 153-161.

Quaternary Proceedings No. 1, 1991 19-25
© Quaternary Research Association, Cambridge

Stratigraphic Resolution and Radiocarbon Dating of Devensian Lateglacial Sediments

J.John Lowe

J.John Lowe, 1991 Stratigraphic Resolution and Radiocarbon Dating of Devensian Lateglacial Sediments, in *Radiocarbon Dating: Recent Applications and Future Potential* (ed. J.J. Lowe). Quaternary Proceedings No. 1, John Wiley & Sons Ltd, Chichester, pp. 19-25.

Abstract

A number of site-specific and systematic errors that can potentially affect radiocarbon measurements obtained from Lateglacial (last glacial-interglacial transition) sediments are reviewed. Attention is given in particular to the influences of three sources of error: (1) inadequate stratigraphic resolution of Lateglacial sediment successions; (2) the complex bio-geochemistry of lake sediments; and (3) temporal variations in atmospheric radiocarbon concentrations during the Lateglacial. An international protocol is necessary in order to establish the extent to which dating and correlation by radiocarbon can be improved.

KEYWORDS: atmospheric radiocarbon variations; site-specific errors; biogeochemistry of lake sediments; international protocol.

Department of Geography, Royal Holloway, University of London, Egham, Surrey, TW20 0EX, U.K.

Introduction

In recent years there has been a growing appreciation of the rapidity and magnitude of climatic changes that occurred during the Devensian Lateglacial (Atkinson *et al.*, 1987) and especially of the influential role of the North Atlantic Polar Front in the climatic history of NW Europe (Ruddiman & McIntyre, 1981; Ruddiman *et al.*, 1977). With the expansion of the deep-sea coring programme, the North Atlantic region has become a major focus of international research interest in view of the evidence for particularly abrupt climatic shifts in the region (e.g. Bard *et al.*, 1987; Labeyrie *et al.*, 1987). Attempts have been made to model oceanographic and climatic changes based on stratigraphic information from the North Atlantic (e.g. Rind *et al.*, 1986) and to explain the mechanisms of ocean temperature change (e.g. Broecker & Denton, 1989). At the same time, evidence for seemingly comparable climatic events has accumulated from other parts of the world, including several parts of North America, as far west as Washington State (Rind *et al.*, 1986), the USSR (Faustova, 1983), New Zealand (Burrows, 1979),South America (Heusser & Rabassa,1987) and the polar regions (ice-core records; see Harvey, 1989). Clearly there is much yet to be learned concerning the overall pattern of climatic shifts during the Lateglacial, and of critical importance will be the ability to date events accurately and precisely.

A major impediment to progress in measuring rates of change during the last glacial-interglacial transition, however, is the lack of a reliable and sufficiently precise dating tool. By comparison with most other geological dating methods, radiocarbon measurements provide a finely-tuned chronology, capable of separating events within a millenium. Within the context of the last glacial-interglacial transition, however, the method has its limitations, as it seldom provides the precision required to measure the very abrupt rates of environmental change that characterise the period. As a result of a range of problems which affect the interpretation of radiocarbon dates, the Lateglacial chronology of climatic and other events, as it is presently understood, tends to be 'blurred'. When considering published radiocarbon dates from NW Europe, for example, it is frequently impossible to separate the effects of dating errors on the one hand from time-transgressive differences between sites on the other (see *e.g.* Gray & Lowe, 1977; Lowe & Walker, 1980,1984).

The development of alternative dating methods, such as tephrachronology, magnetostratigraphy and dendro-chronology, may eventually solve some of the current difficulties of dating and correlation within the Lateglacial, but these are themselves subject to a number of limitations and, moreover, may only have a regional or local application (Becker, in press; Pilcher, in press; Austin and Lowe, 1989; A. Dugdale, unpublished). In any case they rely on radiocarbon measurements as an independent control either for dating or for understanding important global atmospheric flux conditions. Radiocarbon dating will continue to play a dominant role in late Quaternary research for some time to come, and it seems likely that there will continue to be a heavy reliance on radiocarbon dates obtained from lake sediment sequences for regional and inter-regional correlation.

It is therefore imperative that both the accuracy and precision of radiocarbon dating be improved. At a recent IGCP workshop (IGCP-253,*Termination of the Pleistocene*, Oslo-Stockholm-Helsinki, May, 1990) this was identified as the most pressing issue facing the assembled scientific body. Workshop sessions were devoted to a variety of Lateglacial themes, including deglaciation patterns, fluctuations of glaciers, ice-dammed lakes and lake drainage, changes in permafrost conditions, regional climatic curves and vegetational responses to climatic change. The conclusion reached by each specialist group was that dating limitations, especially those associated

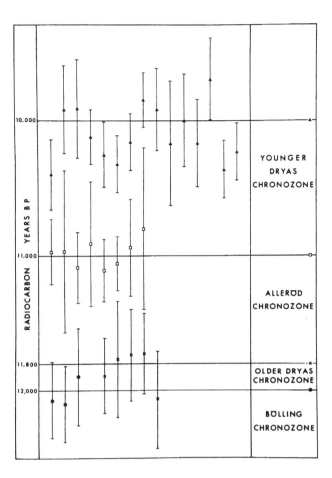

Figure 1 Diagram illustrating the imprecise resolution provided by Lateglacial radiocarbon dates with typical laboratory error ranges (at 2 standard deviations). The dates are of samples obtained from important stratigraphic boundaries at various sites; the spread of the dates and the overlapping error ranges prevent clear resolution of the main stratigraphic intervals, especially of the very short-lived 'Older Dryas' event. (Based on Lowe & Gray, 1980).

with the radiocarbon method, formed the single most important constraint on modelling Lateglacial environmental change.

In this paper it will be argued that significant improvements can still be made in the approach to, and interpretation of, the radiocarbon dating of Lateglacial sediment successions by a more rigorous selection of sites and samples for radiocarbon measurement, thereby providing a more reliable chronology to underpin palaeoenvironmental research programmes. Some sites display a greater degree of stratigraphic resolution than others. This is illustrated, and then the potential for the 'screening' of samples in order to reduce the effects of the more common sources of error will be discussed.

Strategy for an improved radiocarbon chronology of the Devensian Lateglacial

Radiocarbon activity measures obtained from Lateglacial sediment samples are subject to both systematic and site-specific errors. Systematic errors may be caused by the effects of variations in levels of atmospheric radiocarbon during the period 15-9 ka B.P. (Ammann & Lotter, 1989; Bard *et al.*, 1990) or by laboratory bias (Scott *et al.*, this volume). Site-specific

errors may arise from the 'mineral carbon effect', especially in samples of sediment that accumulated in newly-deglaciated terrain (Lowe & Walker, 1980, 1984; Sutherland, 1980; Bjorck & Hakansson, 1982; Lowe *et al.*, 1988), or as a result of the in-washing of older organic carbon (what Olsson terms the 'reservoir effect' - Olsson, 1979, 1986). These errors are compounded where samples have been obtained from sediments that provide a poor stratigraphic resolution of the Lateglacial succession. Since the problems of 'mineral carbon' error and the 'reservoir effect' have already been discussed in detail in the literature, the focus of this paper will be mainly on the other sources of error mentioned above.

1. Influence of stratigraphic resolution

Quoted errors for published dates within the range 15-9 ka radiocarbon years B.P. are at best about +/- 60 but normally greater (at 1 standard deviation). This degree of temporal resolution is often insufficient for detecting short-term events during the Lateglacial (e.g. the 'Older Dryas event') or for testing the synchroneity of abrupt environmental changes (Fig. 1).

It is possible firstly to decrease the laboratory error-ranges of radiocarbon measurements by using larger samples for radiometric assays, and secondly to increase the resolution of events by the use of AMS techniques, but the validity of the derived age measurements depends on the integrity of the geological samples (Lowe *et al.*, 1988; Preece, this volume) and a knowledge of the effects of atmospheric radiocarbon variations (Pilcher, this volume; and discussed below). Initially, however, the most important consideration is the level of stratigraphic resolution offered by the sedimentary sequence from which samples have been obtained.

Improvement in the laboratory precision will depend upon the nature of the samples selected for dating and upon the technical capabilities of the radiocarbon laboratory. Some improvements can be made, however, in reducing the errors associated with inadequate stratigraphic resolution through careful site selection. The point is illustrated in Figure 2 which is based upon published radiocarbon dates obtained from a Lateglacial stratigraphic succession in Scotland. The radiocarbon dates, based on 2cm sediment slices, are indicated with error ranges quoted at 1 s.d.. Using arbitrary dates for the overall stratigraphic interval, and assuming a constant sedimentation rate throughout the period represented by the profile, the time taken for 2 cm of sediment to accumulate at the site (in theory) is about 166 years. Since the s.d. ranges are between 240 and 580, the dates provide inadequate temporal resolution of this particular stratigraphic record.

The calculation of theoretical rates of sedimentation for Lateglacial sediments is difficult, however: highly variable rates would be expected for most sites as a consequence of the unstable environment during the last glacial-interglacial transition. Nevertheless this simplified approach does provide some approximate means of comparing the different levels of chronostratigraphic resolution between sites. Figure 3 shows the estimated intervals of time represented by 1 cm of sediment and of the sediment slices used for radiocarbon dating at each of four published Lateglacial sites from the U.K. using the same basis for calculating theoretical constant sedimentation rates for each site. The beginning and end of the period of the Lateglacial (defined on biostratigraphic criteria) are arbitrarily defined as 13 and 10 ka B.P. respectively. On this basis, Figure 3 illustrates the much greater temporal resolution made possible

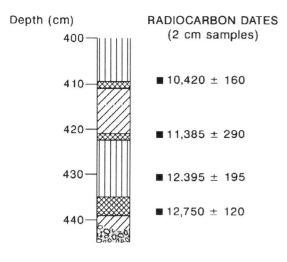

Depth (cm)   RADIOCARBON DATES
              (2 cm samples)

400

410   ■ 10,420 ± 160

420   ■ 11,385 ± 290

430   ■ 12,395 ± 195

               ■ 12,750 ± 120
440

30 cm represents approx 2,500 yrs.

1 cm represents approx 83 yrs.

Each dated (2 cm) sample reps 166 yrs.

Dated age ranges = 240 to 580

(at ONE STANDARD DEVIATION)

Figure 2 Radiocarbon dates obtained from a Lateglacial succession at Tynaspirit, Perthshire, Scotland (from Lowe, 1978). The generalised sediment succession is: gyttja (438-435 and 411-409 cm); fine-detrital peat (435-423 and 409-400 cm); and 'Younger Dryas clay' (423-411 cm). The calculations illustrated are based on arbitrary ages for the base (440 cm) of the Lateglacial organic sediments and of the early Holocene organic sediments (410 cm) of 13k and 10k years respectively and assume a constant sedimentation rate throughout.

by a combination of (a) the use of thinner sediment slices for dating, and (b) the higher sedimentation rates typical of some sites.

In future research, therefore, dating should preferentially be based on samples obtained from sites which have been selected according to resolution potential. Table 1 shows a comparison of Lateglacial sediment accumulation rates calculated for 7 published sites in Wales (S. Lowe, in prep.). This illustrates the very wide-ranging stratigraphic resolution typical of Lateglacial sites in Britain. A data-base of published and unpublished sites in the U.K. showing similar but more comprehensive information is currently being developed (Lowe, J.J. and Walker, M.J.C., unpublished) and forms part of a comprehensive review of the Lateglacial stratigraphy of the British Isles.

Sediment accumulation rate is not the only important variable, however. Much also depends on the chemical nature of the sediments, especially the organic carbon content. Figure 4 shows the measured error ranges for radiocarbon dates published for the same four UK sites used in Figure 3. Whereas in Figure 3 the dated samples from Llyn Gwernan apparently show a much greater *chronostratigraphic* resolution than those of the other sites, it can be seen from Figure 4 that the measured (laboratory) error ranges are significantly greater for the Llyn Gwernan samples than those for the site of Loch an t'Suidhe. This is due to the higher organic carbon content in the

sediments of the latter site, allowing a higher level of precision in the laboratory measurements.

Samples for dating should therefore be selected not only on the basis of stratigraphic resolution, but also from a knowledge of sediment chemistry and in the light of detailed biostratigraphic information.

Table 1 Stratigraphic resolution of radiocarbon-dated Lateglacial sites, Wales: Th = thickness of Lateglacial Interstadial organic sediments (cm); acc/K = sediment accumulation per century (cm).

site	Th	acc/K
Llyn Gwernan	238	11.9
Nant Ffrancon	140	7.0
Llyn Goddionduon	98	4.9
Clogwyngarreg	60	3.0
Low Wray Bay	54	2.7
Glanllynau	30	1.5
Traeth Mawr	30	1.5

2. Biogeochemistry of lake sediments

Further problems in the evaluation of published radiocarbon dates arise from inadequate information on the geochemical and biochemical composition of the dated sediments and from a lack of standardisation in field and laboratory procedures. Lateglacial sediments are highly complex in nature as a result of a number of factors affecting the recruitment of biogenic and clastic material. Very little is known especially about the organic geochemistry of Lateglacial sediments. It is essential that the inorganic and organic geochemistry of sediments be studied more routinely in order to establish their bearing on radiocarbon activity measures, for the following have been established through recent research:

a. radiocarbon measurements obtained from Lateglacial and early Holocene lake sediments are commonly affected by a 'mineral carbon error' (Lowe & Walker, 1980; Sutherland, 1980) caused by minute particles of mineral carbon that can cause significant 'ageing' effects (Nambudiri *et al.*, 1980);

b. the inwashing of older organic carbon detritus is also encountered in Lateglacial deposits (e.g. Lowe & Walker, 1986; Walker & Lowe, 1990), giving rise to a 'reservoir effect' in dated samples (Olsson, 1979, 1986);

c. there is commonly a systematic radiocarbon activity difference between macrofossil cellulose and the host sediments from which the macrofossils have been obtained (Shotton, 1972; Andree *et al.*, 1986; Nelson *et al.*, 1988; Cwynar & Watts, 1989 ; Peteet *et al.*, 1990);

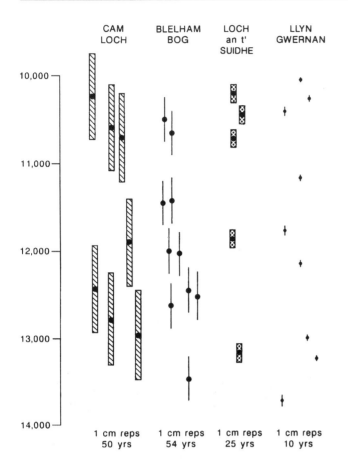

Figure 3 Comparison of length of time taken for 1cm of sediment to accumulate and represented by the sediment samples used for radiocarbon measurements at each of four British Lateglacial sites using the arbitrary boundary dates and method outlined in Figure 2 (and in text). Data from Cam Loch and Blelham Bog obtained from Pennington, 1974; for Loch an t-Suidhe from Walker & Lowe, 1982; for Llyn Gwernan from Lowe (1981) and Lowe & Lowe (1990).

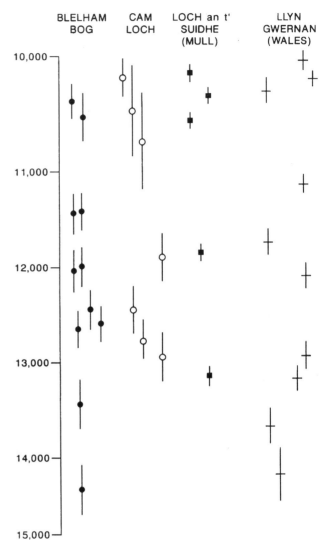

Figure 4 The laboratory error ranges (1 standard deviation) quoted for the dates used in Figure 3; sources as for Fig. 3.

d. different biochemical components from a single sediment sample can give a range of radiocarbon measures (Fowler, 1985; Fowler et al., 1986a, 1986b) as can separate samples from contemporaneous horizons (Lowe et al., 1988).

It is claimed that the more reliable activity measures are obtained from plant macrofossil cellulose. However, this need not always be the case, and future research should continue to focus on the dating of sediments for the following two principal reasons: (a) some plant macrofossil material can be re-cycled or be subject to errors resulting from isotopic fractionation, and (b) many key sites throughout the world lack suitable macrofossil material, so that most correlations will continue to rest on radiocarbon measures obtained from fine-detrital sediments.

More attention should therefore be directed to the investigation of the organic and inorganic chemistry of sediments and the bearing of sediment chemistry on radiocarbon activity measures. Fowler (1985) has shown that it is possible to obtain meaningful AMS activity measures from a range of organic compounds obtained from Lateglacial sediments. In very general terms, fine organic muds consist mostly of humic and fulvic acids (30 to 60% dry weight), cellulose (5-20%) and 'humin' (10-50%). Lipids rarely account for more than 1% dry weight of material, while amino acids are usually less than 1% dry weight (Stevenson, 1982). Nevertheless, with sufficient sample material and careful chemical pretreatment, it is possible to separate these molecular compounds and obtain reasonable measurements of radiocarbon activity using AMS techniques even from the amino-acid component of sediment samples.

AMS measures of the radiocarbon activity of various biochemical components of lake sediments offer much potential for the isolation of errors affecting age estimates. It is possible, for example, to detect the likelihood of 'mineral carbon error' from a knowledge of the geochemistry of samples, from comparisons of radiocarbon dates obtained from macrofossil components compared with bulk sediment samples, or through AMS dating of humin, humic and other chemical 'fractions' of samples (Andree et al., 1986; Lowe et al., 1988; Nelson et al., 1988; Cwynar & Watts, 1989). It is also possible to assess the likelihood of inwashed older organic carbon through AMS dating of humic, lipid and other biochemical components (Lowe et al., 1988) and through analyses of the state of preservation of pollen and spores (Lowe & Walker, 1986; Walker & Lowe, 1990). The heterogeneity of contemporaneous organic compounds can be evaluated through analyses of lipids and other organic derivatives, which potentially can allow the separation of autochthonous and allochthonous components.

There is every possibility, therefore, that the systematic 'screening' of samples using a comprehensive programme of geochemical and biochemical measures, detailed biostratigraphic investigations and AMS measures of the radiocarbon activity of selected sediment components would reduce some of the 'noise' affecting the interpretation and comparison of radiocarbon measurements. The overall objectives would be to identify the most reliable materials and procedures leading to greater consistency in age estimates.

3. Temporal variations in atmospheric ^{14}C concentrations

Detailed radiocarbon stratigraphies of lake sediments from Switzerland suggest that there have been marked variations in atmospheric ^{14}C concentrations during the period 13-9 ka B.P. (Ammann & Lotter, 1989). Two significant 'plateaux' of constant age (occurring at approximately 12.7 and 10.0 ka radiocarbon years B.P.) have been identified in time-depth plots (Fig. 5) which are interpreted as resulting from decreased atmospheric ^{14}C concentrations. If this interpretation is correct, it means that precise ages for Lateglacial events cannot be obtained from radiocarbon measurements that lie within the plateau ranges.

Bard *et al.* (1990) have also recently claimed that there is a systematic and measurable bias in radiocarbon dates for the period 20-10 ka B.P. Their conclusion is based upon the comparison of radiocarbon dates with AMS measures of U-Th ages of coral, the latter considered to be reliable because the

Holocene U-Th measures are in good agreement with tree-ring calibration data. Their findings suggest that radiocarbon measures underestimate the true age by up to several hundred years at 10 k Th years B.P. and by as much as 3.5 kyr at 20 Th ka B.P. (Fig. 6). This trend is also considered to reflect long-term variations in atmospheric ^{14}C concentration.

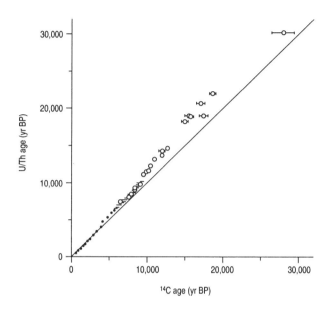

Figure 6 Comparison of the U-Th and ^{14}C ages for the period 0-30 ka B.P. (Based on Bard *et al.*, 1990).

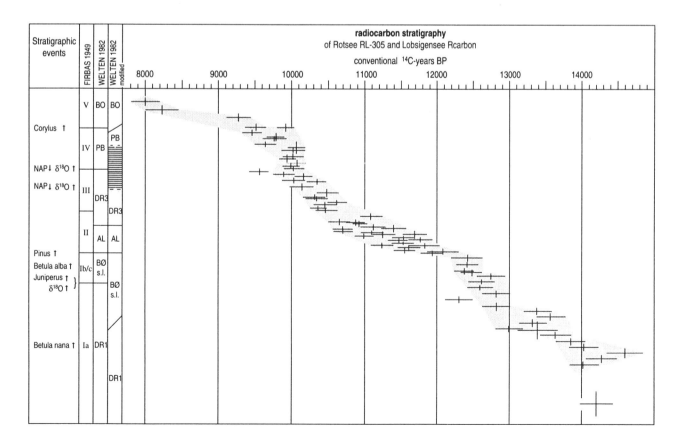

Figure 5 Radiocarbon stratigraphy of Swiss lake sediment successions showing 'plateau' effects at around 12.7 and 10 ka radiocarbon years B.P. (from Ammann & Lotter, 1989).

It is therefore extremely important to examine the validity of these conclusions. The extent to which the radiocarbon chronology of Ammann & Lotter (1989) can be replicated should be a priority objective. If their interpretation is correct, similar time-depth plots should be found in other sites around the world, for such marked variations in atmospheric ^{14}C concentrations as their data imply should be spatially synchronous. This may therefore provide a means of establishing time-parallel correlation of key horizons within Lateglacial sequences. Although absolute ages could not be established by radiocarbon activity measures alone, the degree of synchroneity of environmental changes could be measured through the construction of detailed radiocarbon stratigraphies. A method of calibrating the resulting chronostratigraphies may be achievable through AMS U-Th dating and correlation between the various records may additionally be verified (regionally) through the application of tephrachronological, varve, dendrochronological and palaeomagnetic studies.

Conclusion

In view of the complex web of site-specific (stratigraphic) and methodological problems that affect radiocarbon measurements obtained from Lateglacial sediments, there is a need for an agreed international protocol to minimise errors and to achieve a greater degree of consistency in radiocarbon dates. It would appear that such a protocol should consider the following:

1. the importance of stratigraphic resolution in the assessment of the precision of radiocarbon dates obtained from Lateglacial successions;

2. the importance of establishing the nature of the organic and inorganic geochemistry of sediments in order to establish the reliability of both individual samples and particular chemical fractions of samples, especially where these are used to obtain AMS radiocarbon measures;

3. an examination of the radiocarbon trends established by Ammann & Lotter (1989), to see whether they can be replicated and used as a basis for comparison between sites of high stratigraphic resolution;

4. the use of tephrachronological, varve, dendrochronological and palaeomagnetic studies (where applicable) to underpin regional chronologies established using radiocarbon dates;

5. a further examination of the reliability of the U-Th data obtained by Bard et al. (1990) as a basis for calibrating radiocarbon age measurements within the period 20-10 ka radiocarbon years B.P.

Acknowledgements

The ideas expressed here owe much to extensive discussions over the years with Douglas Harkness, Steve Lowe and Mike Walker, and their valuable contributions are gratefully acknowledged. Mike Walker and Frank Chambers also provided critical comments on earlier drafts of the paper. I am also grateful to the Cartographic Unit of the City of London Polytechnic and to Justin Jacyno (RHBNC) for assistance with the illustrations.

References

AMMANN, B. & LOTTER, A.F. (1989). Late-Glacial radiocarbon- and palynostratigraphy on the Swiss Plateau. Boreas, 18, 109-126.

ANDREE, M. et al. (1986). ^{14}C-dating of plant macrofossils in lake sediment. Radiocarbon, 28, No. 2A, 411-416.

ATKINSON, T.C., BRIFFA, K.R. & COOPE, G.R. (1987). Seasonal temperatures in Britain during the past 22,000 years, reconstructed using beetle remains. Nature, 325, 587-593.

AUSTIN, T.J.F. & LOWE, J.J. (1989). A Late Devensian Lateglacial palaeomagnetic record from Llyn Gwernan, North Wales: comparison with other NW European records. Geophysical Journal International, 99, 699-710.

BARD, E. et al. (1987). Retreat velocity of the North Atlantic polar front during the last deglaciation determined by ^{14}C accelerator mass spectrometry. Nature, 328, 791-794.

BARD, E., HAMELIN, B., FAIRBANKS, R.G. & ZINDLER, A. (1990). Calibration of the ^{14}C timescale over the past 30,000 years using mass spectrometric U-Th ages from Barbados corals. Nature, 345, 405-410.

BECKER, B. (1991). Holocene tree-ring series in the continental European Lowland. In Frenzel, B. (ed.), Proceedings of European Science Foundation Workshop, Arles, 1989, Palaeoklimaforschung Special Volume (in press).

BJORCK, S. & HAKANSSON, S. (1982). Radiocarbon dates from the Late Weichselian lake sediments in South Sweden as a basis for chronostratigraphic subdivision. Boreas, 11, 141-150.

BROECKER, W.S. & DENTON, G.H. (1989). The role of ocean-atmosphere reorganizations in glacial cycles. Geochimica et Cosmochimica Acta, 53, 2465-2501.

BURROWS, C.J. (1979). A chronology for cool-climate episodes in the Southern Hemisphere 12000-1000 yr B.P. Palaeogeography, Palaeoclimatology, Palaeoecology, 27, 287-347.

CWYNAR, L.C. & WATTS, W.A. (1989). Accelerator mass spectrometer ages for Late-Glacial events at Ballybetagh, Ireland. Quaternary Research, 31, 377-380.

DUPLESSY, J.-C. et al. (1981). Deglacial warming of the northeastern Atlantic Ocean: correlation with the palaeo-climatic evolution of the European continent. Palaeogeography, Palaeoclimatology, Palaeoecology, 35, 121-144.

FAUSTOVA, M.A. (1983). Late Pleistocene glaciation of European USSR. In Velichko, A.A. (ed.), Late Quaternary Environments of the Soviet Union, Pergamon Press, 3-12.

FOWLER, A.J. (1985). Radiocarbon dating of lake sediments and peats by accelerator mass spectrometry. Unpublished DPhil Thesis, University of Oxford.

FOWLER, A.J., GILLESPIE, R. & HEDGES, R.E.M. (1986a). Radiocarbon dating of sediments. Radiocarbon, 28, 441-450.

FOWLER, A.J., GILLESPIE, R. & HEDGES, R.E.M. (1986b). Radiocarbon dating of sediments by accelerator mass spectrometry. *Physics of the Earth and Planetary Interiors,* 44, 15-20.

GRAY, J.M. & LOWE, J.J. (1977). The Scottish Lateglacial environment: a synthesis. *In* Gray, J.M. & Lowe, J.J. (eds.), *Studies in the Scottish Lateglacial Environment,* Pergamon, 163-181.

HARVEY, L.D.D. (1989). Modelling the Younger Dryas. *Quaternary Science Reviews,* 8, 137-149.

HEUSSER, C.J. & RABASSA, J. (1987). Cold climate episode of Younger Dryas age in Tierra del Fuego. *Nature,* 328, 609-611.

LABEYRIE, L.D., DUPLESSY, J.-C. & BLANC, P.L. (1987). Variations in mode of formation and temperature of oceanic deep waters over the past 125,000 years. *Nature,* 327, 477-482.

LOWE, J.J. (1978). Radiocarbon-dated Lateglacial and early Flandrian pollen profiles from the Teith Valley, Perthshire, Scotland. *Pollen et Spores,* XX, 367-397.

LOWE, J.J. & GRAY, J.M. (1980). The stratigraphic subdivision of the Lateglacial of NW Europe: a discussion. *In* Lowe, J.J., Gray, J.M. & Robinson, J.E. (eds.), *Studies in the Lateglacial of North-west Europe,* Pergamon Press, Oxford, 157-175.

LOWE, J.J. & LOWE, S. (1989). Interpretation of the pollen stratigraphy of Late Devensian Lateglacial and early Flandrian sediments at Llyn Gwernan, near Cader Idris, North Wales. *New Phytologist,* 113, 391-408.

LOWE, J.J., LOWE, S., FOWLER, A.J., HEDGES, R.E.M. & AUSTIN, T.J.F. (1988). Comparison of accelerator and radiometric radiocarbon measurements obtained from Late Devensian Lateglacial lake sediments from Llyn Gwernan, North Wales, UK. *Boreas,* 17, 355-369.

LOWE, J.J. & WALKER, M.J.C. (1980). Problems associated with radiocarbon dating the close of the Lateglacial in the Rannoch Moor area, Scotland. *In* Lowe, J.J., Gray, J.M. and Robinson, J.E. (eds.), *Studies in the Lateglacial of NW Europe,* Pergamon, 123-137.

LOWE, J.J. & WALKER, M.J.C. (1984). *Reconstructing Quaternary Environments,* Longman, London & New York.

LOWE, J.J. & WALKER, M.J.C. (1986). Lateglacial and early Flandrian environmental history of the Isle of Mull, Inner Hebrides, Scotland. *Transactions of the Royal Society of Edinburgh: Earth Sciences,* 77, 1-20.

LOWE, S. (1981). Radiocarbon dating and stratigraphic resolution in Welsh lateglacial chronology. *Nature,* 293, 210-212.

NAMBUDIRI, E.M.V., TELLER, J.T. & LAST, W.M. (1980). Pre-Quaternary microfossils - a guide to errors in radiocarbon dating. *Geology,* 8, 123-126.

NELSON, R.E., CARTER, L.D. & ROBINSON, S.W. (1988). Anomalous radiocarbon ages from a Holocene detrital organic lens in Alaska and their implications for radiocarbon dating and palaeoenvironmental reconstructions in the Arctic. *Quaternary Research,* 29, 66-71.

OLSSON, I.U. (1979). A warning against radiocarbon dating of samples containing little carbon. *Boreas,* 8, 203-207.

OLSSON, I.U. (1986). Radiocarbon dating. *In* Berglund, B.E. (ed.), *Handbook of Holocene Palaeoecology and Palaeohydrology,* John Wiley, 273-312.

PENNINGTON, W. (1975). A chronostratigraphic comparison of Late-Weichselian and Late-Devensian subdivisions, illustrated by two radiocarbon-dated profiles from western Britain. *Boreas,* 4, 157-171.

PETEET, D.M. *et al.* (1990). Younger Dryas climatic reversal in northeastern USA? AMS ages for an old problem. *Quaternary Research,* 33, 219-230.

PILCHER, J. (1991). Tree-rings in the British Isles and oceanic Europe. *In* Frenzel, B. (ed.), Proceedings of European Science Foundation Workshop, Arles, 1989, *Palaeoklimaforschung* Special Volume (in press).

RIND, D. *et al.* (1986). The impact of cold North Atlantic sea surface temperatures on climate: implications for the Younger Dryas cooling (11-10 ka). *Climate Dynamics,* 1, 3-33.

RUDDIMAN, W.F. & McINTYRE, A. (1981). The North Atlantic Ocean during the last glaciation. *Palaeogeography, Palaeoclimatology, Palaeoecology,* 35, 145-214.

RUDDIMAN, W.F., SANCETTA, C.D. & McINTYRE, A. (1977). Glacial-interglacial response rate of subpolar North Atlantic waters to climatic change: the record in ocean sediments. *Philosophical Transactions of the Royal Society, London,* B 280, 119-142.

SHOTTON, F.W. (1972). An example of hard-water error in radiocarbon dating of vegetable matter. *Nature,* 240, 460-461.

STEVENSON, F.J. (1982). *Humus Chemistry,* John Wiley & Sons, New York.

SUTHERLAND, D.G. (1980). Problems of radiocarbon dating deposits from newly-deglaciated terrain: examples from the Scottish Lateglacial. *In* Lowe, J.J., Gray, J.M. & Robinson, J.E. (eds.), *Studies in the Lateglacial of NW Europe,* Pergamon, 139-149.

WALKER, M.J.C. & LOWE, J.J. (1977). Postglacial environmental history of Rannoch Moor, Scotland. I. Three pollen diagrams from the Kingshouse area. *Journal of Biogeography,* 4, 333-351.

WALKER, M.J.C. & LOWE, J.J. (1982). Lateglacial and early Flandrian chronology of the Isle of Mull, Scotland. *Nature,* 296, 558-561.

WALKER, M.J.C. & LOWE, J.J. (1990). Reconstructing the environmental history of the Last Glacial-Interglacial transition: evidence from the Isle of Skye, Inner Hebrides, Scotland. *Quaternary Science Reviews,* 9, 15-49.

Quaternary Proceedings No. 1, 1991 27-33
© Quaternary Research Association, Cambridge

Radiocarbon Dating for the Quaternary Scientist

J.R. Pilcher

J.R. Pilcher, 1991 Radiocarbon Dating for the Quaternary Scientist, In *Radiocarbon Dating: Recent Applications and Future Potential* (ed. J.J. Lowe). Quaternary Proceedings No. 1, John Wiley & Sons Ltd, Chichester, pp. 27-33.

Abstract

The accuracy of many conventional and AMS radiocarbon dates is not adequate for the sort of questions now being asked in Quaternary studies. The need for, and effects of, radiocarbon calibration are discussed and guide-lines offered for the selection of a laboratory. High precision laboratories and the use of wiggle matching will go a long way to answering the critical questions of rates of change and durations of events in the Holocene.

KEYWORDS: radiocarbon dating; Holocene

Palaeoecology Centre, Queen's University, Belfast, BT7 1NN, Northern Ireland.

Introduction

This review of the radiocarbon dating method will concentrate on recent technical advances and on recent views on the interpretation of radiocarbon results.

For too long Quaternary scientists have accepted radiocarbon dating as an accurate, fool-proof and widely applicable dating method for organic materials of age range between 50,000 years and modern. Few have looked in detail at its limitations or considered the possibility that they might be better off spending their grant money on a crystal ball. The questions being asked by the Quaternary scientist are changing. We know the general trend of Quaternary climate, we know the order in which plants re-entered Europe and North America after the last glaciation and the general trend of Holocene sea level change. The new questions relate to RATES of plant movement, RATES of climate change, RELATIVE ages of events within short time spans and ABSOLUTE ages for comparison with varves and tree-ring dates. It is in these fields that the limitations of radiocarbon dating and also some of the new advances become of great importance.

¹⁴C Dating

The theoretical background to the method and the early history is well covered elsewhere (*eg* Libby, 1965) and up-to-date advice on general aspects of sampling, contamination and calibration are given by Mook and Waterbolk (1985) and by Aitken (1990). There are also a number of special factors of sampling and interpretation of concern to Quaternary scientists; for example Sutherland (1980) discusses problems relating to newly deglaciated terrain and Lowe and Walker (1980) discuss dating the end of the Lateglacial.

The Radiocarbon Laboratory

There are basically three types of laboratory. There are the routine radiocarbon laboratories which measure the radioactivity of the ¹⁴C by gas or scintillation counting, AMS (Accelerator Mass Spectrometry) laboratories which measure the proportion of ¹⁴C atoms directly by mass spectrometry and there are the high precision laboratories. These use similar scintillation or gas counting methods to the routine laboratories, but carry out the measurement to far higher accuracy. We will consider what is meant by accuracy and precision in radiocarbon dating before showing what each type of laboratory is capable of achieving.

Precision of measurement

The precision of measurement depends, in the case of the radioactive counting laboratories, on the total number of radioactive decays of ¹⁴C observed. It follows from this that a bigger sample and/or a longer counting time will give a better precision. However the Poisson statistics of radioactive decay dictate that the standard deviation is approximately the square root of the number of counts (eg 100 counts ±10 (=±10%) or 10,000 counts ±100 (=±1%) or 1,000,000 ±1000 (the same as ± 0.1%)). Note that to get a 10-fold improvement in the standard deviation we have to observe 100 times more ¹⁴C decays: *ie* either increase the sample size or counting time 100-fold. We could observe the sample for 100 times as long, but time is money, so there is a real limitation on long counting times. Increasing the sample size has the same effect, but is limited by the capacity of the machine and very often by the limitations of what is obtainable in the field. Having mentioned standard deviations it is important to note that radiocarbon dates are, by convention, always quoted with ± one standard deviation. Thus there will be about 35% probability that the real date will lie outside this range. For all practical purposes you should double the ± figure in order to obtain the (approx) 95% confidence limits. Unfortunately this will certainly not guarantee that the real date is likely to lie within this range.

Accuracy

Accuracy* of radiocarbon dating is a much more complex matter than precision. Radiocarbon dating is a very complex process and many different parts of the process may affect the accuracy. The fact is that some laboratories are much better at this measurement than others. There are two measures of accuracy relevant to radiocarbon laboratories; there is repeatability which is a measure of how close the answer is likely to be on two identical samples and there is bias which is a measure of how far these answers are from the 'correct' measurement known by other means. In 1982 a test of radiocarbon laboratory repeatability and bias was reported by the International Study Group (1982). The results were poor with many laboratories showing biases of several hundred years and commonly underestimating their measurement errors by a factor of at least 2.5. The laboratories were not identified so the user was left to assume that the laboratory he used was subject to this additional error. As reported in this volume (Scott *et al.*), during 1988 and 1989 a more complex interlaboratory cross-check took place. More aspects of the dating process were checked and a larger range of samples, including wood, peat and shell were tested. The final results were presented in September 1989. Although nearly ten years have passed since the previous interlaboratory tests there is little overall improvement. Some laboratories are clearly very good. They have no significant bias and can repeat measurements within or better than the precision they quote. However there are also laboratories with a bias of 200 years and many that are unable to repeat measurements within a 400 year spread. This disappointing result is made all the more serious for the user because many of the participating laboratories STILL insist on remaining anonymous. This means that without other information we must assume that all those remaining anonymous are AS BAD AS THE WORST. As a general rule the inaccuracies of these laboratories will be covered if you multiply the standard deviation by between 2 and 2.5. For example, if the laboratory quotes a date as 1000 ± 50 we must use this as 1000 ± 125 – then remember to multiply by 2 to get 95% confidence limits *ie* 1000 ± 250 which is a range of half a millennium. For most Holocene problems the date is already known to within this range before radiocarbon dating is undertaken. The recent study suggests that the AMS laboratories in general have a bias of less than 100 years and that their error multiplier is 2 or less. Of all the laboratories, those measuring by scintillation counting seem to be the most erratic having biases up to 250 years and error multipliers of up to x4.

Just to show that this is not an isolated bad example, Baillie (1989) has presented a suite of 38 samples of wood which were radiocarbon dated in several laboratories and subsequently dated by dendrochronology. Their true dates are thus now known, but were not known at the time the radiocarbon measurements were carried out. Of the 38, two samples were out by about 1000 years and 35% of the samples were more than 200 years from the correct date. It is worth stressing that all these measurements were carried out by reputable university or government laboratories. The message is the same, the quoted errors underestimate the true error by x2 or more.

Having looked at the routine and AMS laboratories we must look at the high precision laboratories. The development of high precision laboratories came about in response to a need for greatly improved accuracy by those laboratories involved with radiocarbon calibration (see below). It was necessary for these laboratories to measure the radiocarbon content of known-age samples to an accuracy as good as or better than other laboratories are likely to achieve in the forseeable future. Considerable development work and the most rigorous testing went into the development of procedures to achieve this accuracy (see for example Pearson, 1979). There are only about 5 such laboratories in the world. They include the radiocarbon laboratories of Seattle and Belfast (who have carried out the bulk of the calibration measurements) and also Groningen, Heidelberg and Pretoria who have more recently joined in with the calibration work. Their claims of accuracy are justified in the several publications of the calibration work (*e.g.* Stuiver and Pearson, 1986; Pearson and Stuiver, 1986) where replicate samples measured in different laboratories are compared. Typically these laboratories have a bias never worse than 20 years and a repeatability of about ±10 years. The precision quoted on each date is usually less than ±20 and often as low as ±12 years. Where the error multiplication factor has been calculated it is about 1.2 or less. Thus a high precision laboratory date of 1000 ±15 years specifies the date within a 95% probability range of *ca.* 75 years (15×1.2 = 18, ×2 for 95% limits = 36, ×2 for both sides of the mean = 72). This must be compared with the 500 year range in the routine laboratory example above.

If the high precision laboratories are so good, what is the snag? As well as all the extra care and attention given to all parts of the process, the increase in precision is obtained partly by large sample size and partly by longer counting times. The cost is about double that of routine dates and the requirement for large samples puts a limitation on the contexts from which samples can be obtained. In addition to these limitations the calibration laboratories are still heavily committed to the international calibration work and for the next few years will have only limited capacity for other projects. However it is worth stressing that the accuracy and lack of bias achieved by the high precision laboratories is achievable by any laboratory willing to learn the procedures.

Accurate measurement is still possible even where small sample size precludes the attainment of high precision.

Calibration

[14]C ages as quoted by the laboratory are not calendar ages. They are based on an assumption that the radiocarbon concentration in the atmosphere has remained constant in the past. We now know, by measuring the radiocarbon age of known-age samples, that the atmospheric radiocarbon concentration has varied. It does not vary systematically so there is no theoretical equation that allows us to convert radiocarbon ages to calendar ages. Instead an empirical calibration has been established using wood whose age is known exactly by dendrochronology. So far this calibration extends back about 9000 years.

Quaternary scientists often claim that calibration does not matter as all dates are still comparable one with another. This may have been true for the early studies of the Post-glacial, but the new wave of Quaternary studies interested in rates of change and absolute timescales requires that the dates are calibrated. Two hypothetical examples will show what effect calibration can have. Let us assume that the arrival of a particular pollen type at two sites is dated to 4100 ± 50 BP and 4080 ± 50. Within their precisions these dates are clearly

---

* accuracy and precision: a precise watch that tells the time to the nearest second may still be inaccurate by 10 minutes. An imprecise watch with no second hand may still be accurate and tell exactly the correct time.

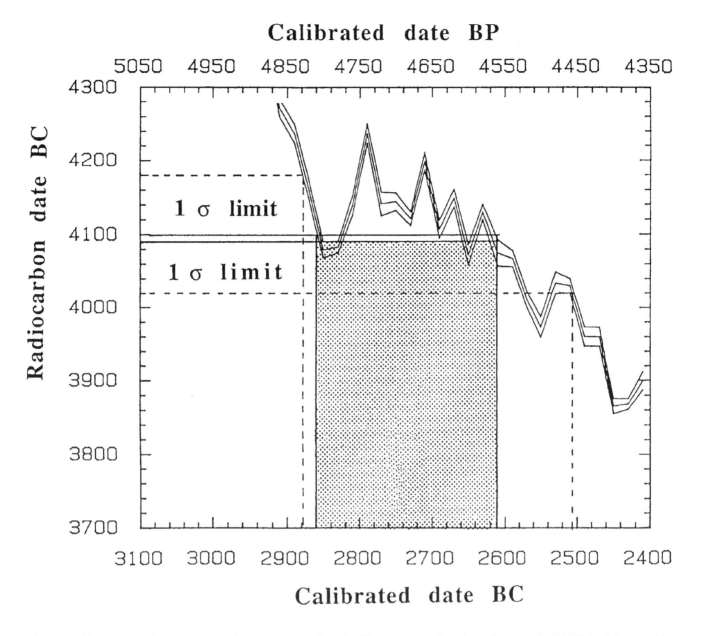

Figure 1 Calibration curve from Pearson *et al.* 1986 showing effect of calibrating radiocarbon dates of 4100 and 4090 BP. Real dates could be anywhere in range 4830 and 4470 and thus one or other sample could be at least 300 years older than the other rather than contemporary as suggested by the raw dates.

inseparable and the analyst assumes that the arrival is synchronous. Look now at the section of calibration curve in Fig 1. The dates with their ranges are marked on the radiocarbon (vertical) axis. Reading off from the true age axis, we can see that the samples could date from anywhere between 4470 and 4830. Thus either one or other of the samples could be significantly older than the other. We can no longer conclude that they are synchronous.

As a second example let us assume that a pollen analyst wants to look for the effects on the vegetation of the dust veil from the Santorini (Thera) eruption. Recent evidence points to this event being in the year 1628 BC (Baillie, 1990). This date is derived from dendrochronology and is thus a calendrical date. The pollen diagrams will be dated by radiocarbon dating. If the pollen analyst looks at about 1628 BC on the radiocarbon scale he will not find the eruption. A look at the calibration curve in Fig 2 shows that a real age of 1628 BC has a

radiocarbon concentration equivalent to a date of 1375 BC and this is where the pollen analyst must search in his pollen record.

Calibration curves back to 7000 years are published (Stuiver and Pearson, 1986; Pearson and Stuiver, 1986; Pearson *et al.*, 1986, Pearson *et al.* in press) and the continuation back to 9000 years has been measured and will be available soon. Beyond that there is thought to be enough dated wood to extend back to about 10,000 years. To extend the calibration beyond that will require a large international effort on the construction of tree-ring chronologies, before the calibration measurements can continue.

Calibration beyond the present limits of the calibration curve.

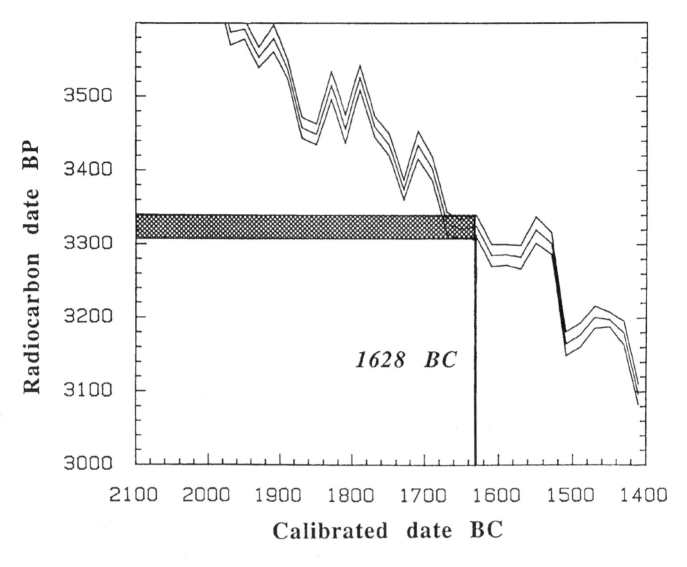

Figure 2 Calibration curve from Pearson *et al.* 1986 showing location of radiocarbon age equivalent to a real date of 1628 BC. An actual radiocarbon measurement will only be as close to this age as the precision and accuracy of the laboratory allow.

The construction of a calibration curve depends on the existence of known age samples, normally from tree-ring dated wood. The present older limit of tree-ring chronologies is about 9000 years with some sections of wood running back to about 10,000 years. Unfortunately, one of the most interesting time-spans for the Quaternary scientist is the Lateglacial – say from 14,000 to 10,000 years ago on the radiocarbon timescale. This was a time of rapid climate change and is of topical interest as a test bed for ideas of rates of climate change in the high CO_2 world of the next 100 years. Clearly to answer the sorts of questions the climatologists are asking will require an accurate measure of RATES of change and thus accurate timescales. At present radiocarbon dating cannot provide this accurate timescale. There is no problem in measuring the radiocarbon content, but without a calibration curve there is no way of interpreting these measurements as dates. From a theoretical viewpoint we might expect major fluctuations in the radiocarbon content of the atmosphere as the ice sheets retreated. Old CO_2 trapped in the ice sheets may have been released as the ice melted and also the solubility of CO_2 in water is temperature dependent. Both these changes and changes in the earth's biomass at the start of the Holocene will have affected the balance of the global carbon cycle and inevitably the proportions of ^{14}C. Fig 3 shows the current best estimate of the changes in ^{14}C concentration of the atmosphere over the 14,000 to 10,000 period. Clearly it is of the highest priority that tree-ring dated wood in the age range be found. Possible sources of wood are known in California, Florida, South of the Alps in Europe, in Tasmania and in New Zealand and work is in progress to study these.

Until this section of calibration is complete it would be wise to make NO interpretation of RATES of change or ABSOLUTE ages for the Lateglacial based on radiocarbon dating.

Radiocarbon dating before 14,000

All the uncertainties that relate to the Lateglacial also apply to the full glacial and interstadials before it. At present we have no idea what way the calibration curve will go. In addition, beyond 15 ka. there is a fairly small amount of ^{14}C left in the sample. The amount decreases by a half for every 5.5 k years, thus by 30 ka there is only 1% and by 50 ka there is only 0.1%.

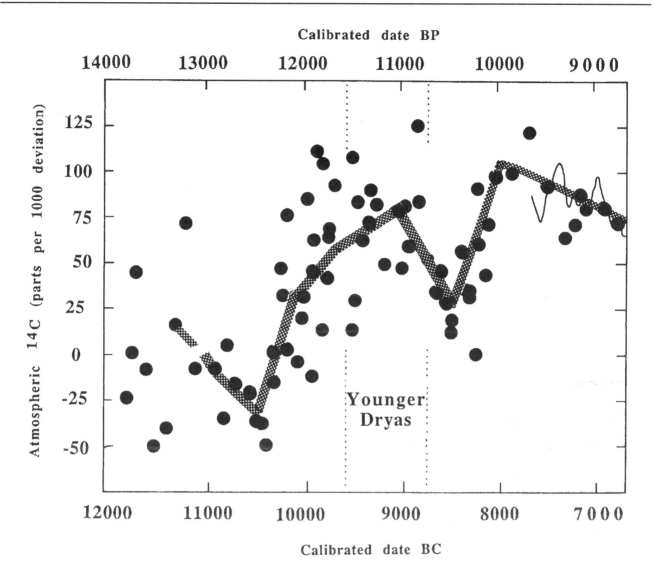

Figure 3 Carbon-14 activity in the atmosphere as estimated by Stuiver *et al.* from sources based on wood, lake and varve samples. The deviation of ±100 per mil represents almost 1000 years. These results suggest fluctuations of about 500 years at the start of the post glacial and of 1000 years during the Alleröd (Stuiver *et al.* 1991).

As the modern ¹⁴C concentration is only 1 part in 10¹², these concentrations are positively homeopathic! The detection limits for most laboratories is about 45 ka. This limit is set by the background radiation detected by the machine. This is related to machine design and to location. Laboratories some distance below ground are protected from cosmic rays and have a lower background. The AMS laboratories have the theoretical capability of dating back to about 65ka, but at present they also have problems of background related to machine design. Current work on high accuracy Uranium/Thorium series dating by mass spectrometry may eventually provide a calibration curve for those parts of the ¹⁴C range not covered by tree-rings. Bard *et al.* 1990 describe U/Th dating of corals, and the construction of an outine calibration back to 30 ka, but the carbon content of corals will only allow AMS dating giving a poor precision to the ¹⁴C axis of the calibration. At present the real limitation on dating of samples beyond 30 ka is contamination both in the field and in the laboratory.

Wiggle matching
– the ultimate radiocarbon date

Wiggle matching is a state-of-the-art technique for obtaining a very accurate estimate of calendar age by using the short-term wiggles or variations in the calibration curve. It is necessary to have a series of samples whose real spacing in time is known. The ideal is a sample of wood of about 100 annual rings. This is divided into 5x 20-year blocks and each measured by high precision radiocarbon dating. The five samples form a short section of calibration curve that can be matched against the full calibration curve. Depending on the sharpness of detail in the calibration curve at the appropriate date, the section can often be matched to within a few years on the calendar scale (Pearson, 1986). While a wood sample is ideal for wiggle matching, Clymo *et al.* (1990) have demonstrated that it is possible to use the technique for peat and lake sediments where the deposition rate can be roughly estimated. This technique is expensive – requiring at least 5 high precision dates per wiggle match (Clymo *et al.* used 7 dates) – but is capable of achieving dates of accuracy comparable with dendrochronology and dates of sufficient resolution to answer for the first time questions relating to rapid rates of change in climate and vegetation.

The future of radiocarbon dating

Clearly now that the high precision laboratories have demonstrated what is possible there will eventually be a move towards better accuracy in all laboratories. This will happen sooner rather than later if users apply sufficient pressure. There are signs that major changes are happening. A new quality assurance protocol is being developed by the radiocarbon community and this is being backed up by a range of known-age samples which will be supplied by the International Atomic Energy Agency, Vienna.

Recommendations to ¹⁴C users

1. Consider whether the problem you wish to solve is amenable to radiocarbon dating. If it relates to small time differences or to rates of change beyond the present calibration, or is likely to be older than 50ka, the method is not appropriate. Dating such samples is not only a waste of money, it silts up the literature with useless information.

2. Consider whether the material you submit to the laboratory was growing at a time contemporary with the event you wish to date. Remember wood and charcoal can persist for a long time, roots can penetrate deposits, sediment can be re-worked, *etc*. If you can be sure of the stratigraphic validity of only one component such as the seeds or pollen, then consider using AMS dating on that fraction rather than dating the whole sediment (Andree *et al.* 1986).

3a. Consider whether the cost of high precision dating is justified. If it is, then discuss the project with a high-precision laboratory and design a sampling strategy to obtain large enough high-integrity samples.

3b. Could the cost of a wiggle-matching exercise on wood or sediment be justified in terms of the greatly improved precision now possible?

4. For any samples that are more than 20ka, involve the radiocarbon laboratory at the project design stage. Preferably get the laboratory personnel to take the samples in the field and be responsible for them at all times from then to completed measurement. Under no circumstances bring such samples into a University science department where it is likely that tracer experiments are carried out (all Biology departments! - see Pilcher, 1990).

5. INTERROGATE YOUR ¹⁴C LABORATORY. It is your duty to science and to your funding agency to get value for money. Value for money means getting useful accurate dates rather than picking the cheapest dates. Ask your chosen laboratory how they have fared in recent interlaboratory tests; ask to see the results. If they will not show you the results assume they are bad and change your lab. If they have not taken part (perhaps for good reasons), ask to SEE other evidence of an ONGOING QUALITY ASSURANCE PROGRAM. If you are not satisfied, change your lab. Do not let funding agencies force you into using

one particular laboratory if you can demonstrate that the quality assurance is absent or inadequate. Finally remember that the results of an interlaboratory test series or an internal quality assurance programme are likely to be the BEST that the laboratory is capable of.

6. Ask the radiocarbon laboratory to carry out the calibration for you; it is an integral part of the measurement and the laboratory should be in the best position to estimate their own errors and make a valid calibration.

References

AITKEN, M.J. (1990). *Science-based dating in Archaeology.* Longman, London.

ANDREE, M., OESCHGER, H., SIEGENTHALER, U., REISEN, T., MOELL, M., AMMAN, B. & TOBOLSKI, K. (1986). ¹⁴C dating of plant macrofossils in lake sediment. *Radiocarbon* 28, 411-416.

BAILLIE, M.G.L. (1989). Checking back on an assemblage of published radiocarbon dates. *Radiocarbon* in press.

BAILLIE, M.G.L. (1990). Checking back on an assemblage of published radiocarbon dates. *Radiocarbon,* 32, 361-366.

BAILLIE, M.G.L. & MUNRO, M.A.R. (1988). Irish tree rings, Santorini and volcanic dust veils. *Nature,* 332, 344-346.

BARD, E., HAMELIN, B., FAIRBANKS, R.G. & ZINDLER, A. (1990). Calibration of the ¹⁴C timescale over the past 30,000 years using mass spectrometric U-Th ages from Barbados corals. *Nature* 345, 405-410.

CLYMO, R.S., OLDFIELD, F. APPLEBY, P.G., PEARSON, G.W., RATNESAR, P. & RICHARDSON, N. (1990). The record of atmospheric deposition on a rainwater-dependent peatland. *Philosophical Transactions of the Royal Society, London, B* 327, 331-338.

CWYNAR, LES C. & WATTS W.A. (1989). Accelerator-Mass Spectrometer ages for Late-glacial events at Ballybetagh, Ireland. *Quaternary Research* 31, 377-380.

EDWARDS, R.L., CHEN, J.H. & WASSERBURG, G.J. (1987). Precise timing of the last interglacial period from mass spectrometric determination of thorium-230 in corals. *Science* 236, 1547-1553.

INTERNATIONAL STUDY GROUP (1982). An interlaboratory comparison of radiocarbon measurements in tree-rings. *Nature,* 298, 619-623.

LIBBY, W.F. (1965). *Radiocarbon dating* (2nd edition), University of Chicago Press, Chicago.

LOWE, J.J. & WALKER, M.J.C. (1980). Problems associated with radiocarbon dating the close of the Lateglacial period in the Rannoch Moor area, Scotland. in *Studies in the Lateglacial of North-West Europe* Ed. by J.J. Lowe, J.M. Gray and J.E. Robinson, Pergamon Press, Oxford, pp123-137.

MOOK, W.G. & WATERBOLK, H.T. (1985). *Handbooks for Archaeologists No 3: Radiocarbon dating.* European Science

Foundation, Strasbourg.

PEARSON, G.W. (1979). Precise ^{14}C measurement by liquid scintillation counting. *Radiocarbon* 21, 1-21.

PEARSON, G.W. (1986). Precise calendrical dating of known growth period samples using a 'curve fitting' technique. *Radiocarbon* 28, 292-299.

PEARSON, G.W., PILCHER, J.R., BAILLIE, M.G.L., CORBETT, D.M. & QUA, F. (1986). High precision ^{14}C measurement of Irish oaks to show the natural ^{14}C variations from AD 1840-5210 BC. *Radiocarbon* 30, 911-934.

PEARSON, G.W., BECKER, B. & QUA, F. (in press). High precision ^{14}C measurements of German oaks to show the natural ^{14}C variations from 6102 to 5090 BC. *Radiocarbon* 31 (3) in press.

PEARSON, G.W. & STUIVER, M. (1986). High precision calibration of the radiocarbon timescale, 5000-2500 BC.

Radiocarbon 28, 839-862.

PILCHER, J.R. (1990). *Radiocarbon dating - a user's guide.* Quaternary Research Association Dating Manual Ed. by P.L. Smart (in press)

STUIVER, M. & PEARSON G.W. (1986). High precision calibration of the radiocarbon timescale, AD 1950-500 BC. *Radiocarbon* 28, 805-838.

STUIVER, M., BRAZIUNAS, T.F., BECKER, B. & KROMER, B. (1991). Climatic, solar, oceanic and geomagnetic influences on Late Glacial and Holocene Atmospheric ^{14}C/^{12}C change. *Quaternary Research* 35, 1-24.

SUTHERLAND, D.G. (1980). Problems of radiocarbon dating deposits from newly deglaciated terrain: examples from the Scottish Lateglacial. In *Studies in the Lateglacial of North-West Europe* Ed. by J.J. Lowe, J.M. Gray and J.E. Robinson, Pergamon Press, Oxford, pp123-137.

Quaternary Proceedings No. 1, 1991 35-43
© Quaternary Research Association, Cambridge

The Devensian Lateglacial Carbon Isotope Record from Llanilid, South Wales

D.D. Harkness and M.J.C. Walker

D.D. Harkness and M.J.C. Walker, 1991 The Devensian Lateglacial Carbon Isotope Record from Llanilid, South Wales, In *Radiocarbon Dating: Recent Applications and Future Potential* (ed. J.J. Lowe). Quaternary Proceedings No. 1, John Wiley & Sons Ltd, Chichester, pp.35-43.

Abstract

Some typical problems encountered with the ^{14}C record in Lateglacial sediments, and a first approach to their solution, have been described and discussed in relation to a Devensian Lateglacial and early Flandrian sedimentary succession exposed by opencast coal mining near Llanilid, South Wales (Walker & Harkness, 1990). In this paper, a more detailed account is provided of the rationale and procedures adopted in the selection of component organic material for ^{14}C age measurement, and the significance of this approach is evaluated in terms of the eventual precision and accuracy of the resulting radiocarbon chronology. Variations in organic productivity (wt % Carbon) and stable isotope enrichment (δ^{13}C) measured at regular intervals throughout the c. 1.8 m of sediment that accumulated during the period 13,200 to 9400 years BP are also considered.

KEYWORDS: radiocarbon rating; chemical fractions; ^{13}C enrichment

D.D. Harkness, NERC Radiocarbon Laboratory, East Kilbride, Glasgow G75 0QU

M.J.C. Walker, Department of Geography, St David's University College, University of Wales, Lampeter, Dyfed SA48 7ED, Wales, U.K.

Introduction

It is becoming increasingly apparent that carbon isotope geochemistry has a central role to play in the search for a better understanding of the controlling influences and consequences of global environmental change. This is, perhaps, most clearly reflected in the dependence on radiocarbon dating for the establishment of a timescale of climatic events over the past 50,000 years. However, the ^{14}C chronology is by no means absolute, and the reliability of dates depends upon the geological integrity of the organic material used for age measurement. Lateglacial sediments, which accumulated during a period of rapid and often abrupt climatic/environmental change, pose a particular challenge for chronological reconstruction; first in the identification and isolation of unambiguously representative carbon for isotopic analysis, and secondly in the objective interpretation of the resulting age measurements.

The development of a ^{14}C timescale from the sequence of Lateglacial and early Flandrian sediments exposed at Llanilid, South Wales (Walker & Harkness, 1990), served to highlight the problems encountered in the dating of events at the last glacial/present interglacial transition. It also clearly identified research priorities in ^{14}C dating that are now considered essential if present levels of instrumental capability are to be translated into a more precise timescale of environmental and/or climatic change.

The Site

The site is located in an opencast coal-working immediately to the east of the village of Llanilid (Nat. Grid Ref. SS 984 818) near Bridgend in Mid Glamorgan, South Wales. It lies at an altitude of c. 60 m OD in an area of ice stagnation landforms that developed immediately within the Late Devensian ice limit which lay some 3 km to the west and south of the site. The underlying geology consists of Carboniferous shales of the Lower Coal Measures. Mining operations during the spring of 1986 truncated one of the many kettle hole basins in the area to expose c. 1.8 m of limnic, semi-terrestrial and terrestrial deposits of Lateglacial and early Flandrian age (Figure 1). Unfortunately, the section no longer exists as the working face of the opencast mine moved through this part of the site later in the year.

At the time of sampling, the section consisted of a lower suite of silty clays (0-4 cm above basal datum), overlain first by silty organic muds (4-15 cm) and then by c. 85 cm of fine-grained organic muds. These were succeeded around 98 cm by a thin lens of clay gyttja and from 100-122 cm by a unit of silt/clay. This upper minerogenic suite was overlain, successively, by organic muds (122-129 cm), fibrous peats (129-138 cm), detritus mud (138-147 cm) and amorphous peats (147-180 cm). Eight sequential monoliths, each measuring approximately 25 by 25 cm and 20 cm deep were cut from the section for laboratory analysis.

Research Strategy

As already described in Walker & Harkness (1990), the sedimentary sequence from the Llanilid kettle hole basin yielded a comprehensive palynological record reflecting

Figure 1 Lateglacial sediment succession exposed in 1986 at Llanilid, S. Wales (Photo M.J.C. Walker).

LLANILID, SOUTH WALES
Percentage Pollen Diagram [principal terrestrial taxa only]

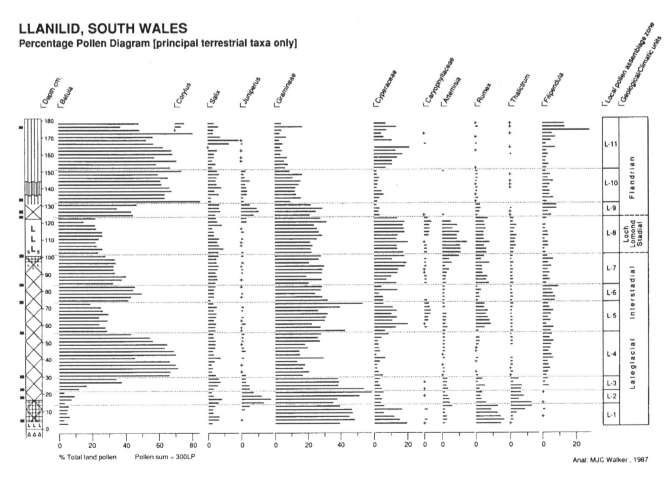

Figure 2 Percentage pollen diagram from the Llanilid profile, showing the principal pollen taxa and the horizons selected for ^{14}C dating (■). The depth scale relates to an arbitrary datum at the base of the section (from Walker and Harkness, 1990).

vegetational changes during the last glacial/interglacial transition (Figure 2). This detailed biostratigraphic framework also provided an ideal opportunity, due to the virtually unlimited availability of organic material, to attempt a comprehensive and stratigraphically precise reconstruction of a ^{14}C timescale for sediment accumulation. In particular, with such large amounts of material available, it was hoped that a solution (or at least a partial solution) could be found to some of the problems commonly encountered in radiocarbon dating of Late Quaternary limnic sediments. These include the possible, frequently masked, presence of organic carbon fixed under hard-water conditions, and contamination by reworked organic residues by mineral carbon and/or younger carbon emplaced by rootlet decay or water transport.

In the absence of extractable macrofossil material, the aim was to establish the ^{14}C ages of the component organic content of specific biostratigraphic horizons. The experimental approach centred on the quantitative separation and independent dating of alkali soluble (humic) and alkali insoluble (humin) components in the acid washed organic detritus.

As the ^{14}C dating programme progressed, an apparent correlation began to emerge between the age/climatic pattern and the stable (δ^{13}C) isotope enrichment values routinely monitored as part of the radiometric dating procedure. It was therefore decided to measure the relative organic productivity and corresponding stable isotope enrichment (δ^{13}C) signatures independently and in more detail. The results of this aspect of the dating programme are discussed in the second part of the paper.

Differential Carbon Content and the [14]C Record

Sampling strategy and chemical subdivision

As substantial monoliths were available (as opposed to sediment cores), large amounts of material could be taken in relatively thin lenses from each sampling horizon. Twelve levels were selected for ^{14}C dating on the basis of changes in the pollen stratigraphy and, in each case, a slice of material 1.5-2.0 cm in thickness was cut from that part of the monolith that had been set aside for dating purposes. The dry weight of the samples ranged from c. 210 to 450 g.

Samples returned to the laboratory were dried to constant weight in a vacuum oven at temperatures not exceeding 40°C. The purpose was to provide a common reference composition for the calculation of component carbon contents (i.e. as weight % Carbon in raw dried sediment). Samples of this material (c. 100-200 g) were subjected to an identical procedure (Figure 3) to separate and recover the component 'humic' and 'humin' carbon intended for ^{14}C age measurement. Initially, the samples were given the pretreatment normally applied to remove the acid soluble (fulvic) material from organic material prior to ^{14}C dating. This digestion in dilute hydrochloric acid also served to ensure removal of any carbonate residues present in the mineral sediment matrix. The subsequent separation of the acid washed organic material into its 'humic' (alkali soluble) and 'humin' (alkali insoluble) components was based on exhaustive extraction with dilute potassium hydroxide solution. Care was taken during all chemical pretreatment and separation stages to obtain a quantitative recovery of the defined organic products. This quantitative procedure was essential to allow a meaningful comparison of different carbon contents and to enable the calculation of weighted mean age values that would effectively represent the total acid washed

carbon in each of the bulk sediment samples.

Natural partitioning of the organic carbon residues

Carbon contents were determined for the raw (oven dried) sediment and for its 'humic' and 'humin' extracts. Aliquots each containing c. 0.5 g of sample material were combusted in an atmosphere of pure oxygen and the resultant carbon dioxide measured in a calibrated standard volume incorporated in the semi-micro oxidation apparatus. Components of the total carbon inventory of each sample are shown in Table 1. The amount of acid soluble (fulvic) organic carbon in each sample was calculated by difference from the measured values of total carbon content.

Table 1 Content and distribution of organic carbon in the Llanilid sediment selected for ^{14}C age measurement.

Sample Depth (cm)	Wt% carbon in raw sediment			
	Humic	Humin	Fulvic	Total
176	22.2	11.3	13.1	46.6
133	23.0	9.5	12.9	45.4
126	19.1	8.8	5.1	33.0
123	13.0	6.7	1.9	21.6
100	6.7	5.0	3.5	15.2
83	12.9	4.8	13.4	31.1
73	14.2	7.1	10.7	32.0
55	24.5	11.4	11.4	47.3
30	19.6	9.5	12.1	41.2
22	8.2	7.0	4.8	20.0
18	8.7	6.7	3.8	18.2
5	1.1	1.3	2.4	4.8

A significant feature of the measured variations in the defined organic carbon components (Figure 4) is their close correspondence with the palynological and lithostratigraphic evidence. There is clear evidence for a marked decrease in organic productivity during implied cold or cooler phases. However, the pattern also reflects a climatic sensitivity in the occurrence of the more labile 'fulvic' and 'humic' forms in relation to the chemically inert 'humin' component of the organic residues. The ratio of 'humic' to 'humin' carbon appears close to unity in detritus laid down during stadial conditions, increasing by a factor of two or more with climatic amelioration.

Calculation and comparison of ^{14}C age values

The conventional ^{14}C ages measured for 'humic' and 'humin' carbon in the Llanilid profile are shown in Table 2, together with the corresponding weighted mean values calculated to represent the total acid insoluble component of the organic detritus. These latter age values are the likely values that would have been obtained by conventional pretreatments of bulk samples from this site (i.e. digestion in dilute mineral acid followed by washing to neutral pH). They were calculated, to ± 50 year confidence, using the relationship:

$$D_T = D_c f_c + D_n f_n$$

where '**D**' denotes ^{14}C isotopic enrichment (‰) and '**f**' the relative

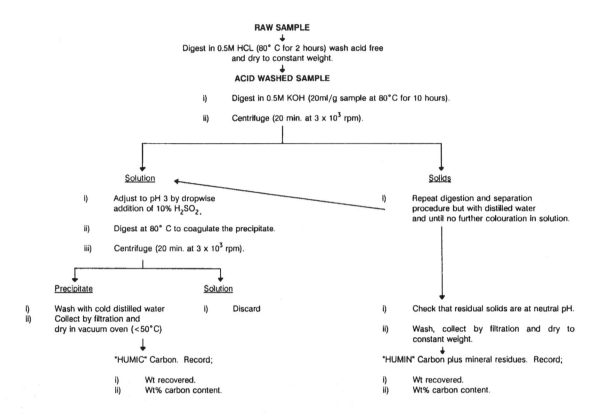

Figure 3 Schematic of chemical procedures used for the initial pretreatment of raw sediment and the quatitative recovery of 'humic' and 'humin' carbon from the acid-insoluble organic residues.

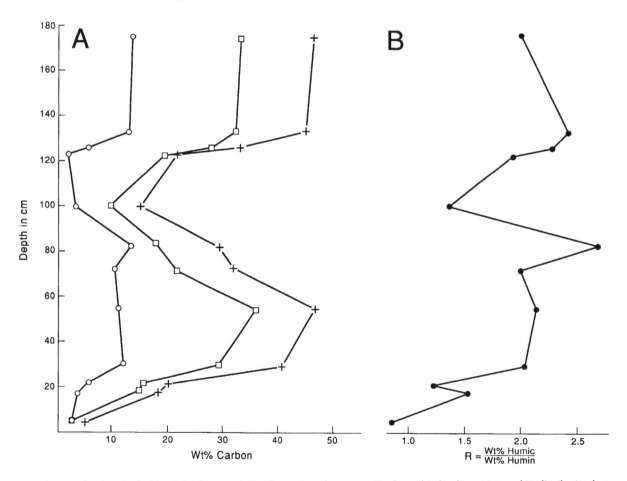

Figure 4 Distribution of carbon in the Llanilid sediments. A. Total organic carbon; as wt % of raw dried sediment (+); and its distribution between 'acid-soluble' (O) and 'acid-insoluble' (□) components. B. The ratio (R) of 'humic' to 'humin' carbon in the total 'acid-insoluble' detritus.

contribution from 'humic' (c) and 'humin' (n) carbon to the total (T) pool of acid-insoluble carbon. The corresponding conventional radiocarbon age (A) is then defined as:

$$A = 8033 \log e \left[1 - \frac{D_T}{1000} \right] \text{ years BP}$$

Table 2 Measured and calculated (weighted mean) ^{14}C age values from Llanilid sediments

Sample Horizon (depth cm)	Conventional ^{14}C Ages (years BP $\pm$ 1 σ)		
	Humic Carbon	Humin Carbon	Weighted mean
176	9140 $\pm$ 65	9320 $\pm$ 60	9200 $\pm$ 45
133	9015 $\pm$ 65	9570 $\pm$ 60	9170 $\pm$ 45
126	9355 $\pm$ 65	9850 $\pm$ 65	9510 $\pm$ 45
123	9410 $\pm$ 65	9920 $\pm$ 65	9580 $\pm$ 50
100	11,080 $\pm$ 70	11,160 $\pm$ 70	11,110 $\pm$ 50
83	11,060 $\pm$ 70	11,470 $\pm$ 80	11,175 $\pm$ 55
73	11,080 $\pm$ 70	11,300 $\pm$ 70	11,150 $\pm$ 50
55	11,170 $\pm$ 70	11,410 $\pm$ 70	11,245 $\pm$ 50
30	11,710 $\pm$ 75	11,655 $\pm$ 70	11,690 $\pm$ 55
22	12,140 $\pm$ 70	12,255 $\pm$ 70	12,190 $\pm$ 50
18	12,380 $\pm$ 70	12,495 $\pm$ 70	12,420 $\pm$ 50
5	13,200 $\pm$ 75	14,200 $\pm$ 75	13,685 $\pm$ 55

A significant, often considerable difference in 'humic' and 'humin' ages is apparent for several horizons in the Llanilid profile (Table 2) and with a general tendency for the 'humin' component to record the older age values. In this instance it is instructive to compare the component ^{14}C age pairs using a 2σ (95% probability) criterion for statistical concurrence *i.e.*

For age agreement $\Delta \leq 2 \sqrt{(\sigma_n^2 - \sigma_c^2)}$

where 'Δ' equals the numerical difference between the median age values for the 'humic' and 'humin' carbon and σ represents the standard deviation ($\pm$ term) of the relative ages. This analysis shows (Table 3) that for eight of the twelve horizons from the Llanilid profile, the age differences between the 'humic' and 'humin' fractions are greater than the 2σ level achieved in the routine ^{14}C dating. The preselection and/or unambiguous identification of material for ^{14}C measurement is therefore a significant factor in such instances.

With the exception of the basal sample from the profile, where the 'humic' date may well be the most reliable estimate of age, the 'humin' age values are considered to provide the least biogeochemically distorted timescale for Lateglacial and early Flandrian deposition at Llanilid (Walker & Harkness, 1990). The younging effect from 'humic' carbon (which was considered to be the principal source of error in the Llanilid series) is most noticeable in the early Flandrian part of the profile where it becomes significant even at relatively crude (*c.* $\pm$ 150 year) levels of analytical precision (Table 3). In the Lateglacial sediments the component age range is much less,

but nevertheless it can still be significant where overall analytical precision equates to better than $\pm$ 80 years (Table 3).

The implications for ^{14}C dating strategy

The Llanilid ^{14}C data set has important implications for present and future attempts to establish the timing of events during the Lateglacial to Flandrian transition. In any ^{14}C dating exercise, the prime consideration is that of achieving both accuracy and analytical precision *i.e.* the ability to resolve adjacent events on the radiometric timescale (Scott *et al.,* this volume). All too often, however, the significance of the age result is discussed solely on the basis of the quoted analytical precision ($\pm$ term).

Table 3 Statistical comparison of age values measured from component humic and humin carbon in the Llanilid sediments

Sample Horizon (depth cm)	Median Age Difference (Δ)	Error Comparator	Statistical Concordance
176	180	178	Outwith 2 σ
133	555	178	Outwith 6 σ
126	495	178	Outwith 5 σ
123	510	184	Outwith 5 σ
100	80	198	1σ agreement
83	410	213	Outwith 3 σ
73	220	198	Outwith 2 σ
55	240	198	Outwith 2 σ
30	-55	205	1 σ agreement
22	115	198	2 σ agreement
18	115	198	2 σ agreement
5	1000	212	Outwith 9 σ

Δ = (Humin-Humic) median age value

In most, if not all instances, this attitude represents undue optimism. The outcome of the Llanilid dating excercise points clearly to the fact that the true level of confidence that can be ascribed in chronological interpretation is much more likely to be constrained by sample integrity i.e. the question of whether the organic residues selected for ^{14}C measurement are uncontaminated by carbon from younger or older sources. It follows, therefore, that although accuracy will always remain the paramount objective in natural ^{14}C measurement, there is little point in striving for a level of analytical precision that exceeds the degree of uncertainty arising from the nature and context of the available organic material. Of the many types of sample encountered in ^{14}C dating, few can be considered to possess the compositional and/or contextual quality that matches the level of age discrimination implied by 'high precision' ^{14}C analyses (*i.e.* $\pm$ 20 years or better). Notable exceptions are wood recovered from well-defined dendrochronological or historical sequences, and similar growth increments in contamination-free corals. As is discussed elsewhere in these proceedings, high-precision dating of material from each of these sources has an invaluable role to play in establishing the calibration indices necessary to convert conventional ^{14}C ages to calendar years (Pilcher, this volume).

One solution to the problem of compositional and/or contextual doubts about available sample material lies in the recovery of identifiable macrofossils for ^{14}C measurement by accelerator mass spectrometry (Hedges, this volume). However, for routine applications, it would seem that the AMS method has yet to match the precision of radiometric dating in general, let alone achieve the 'high-precision' accolade.

Conventional radiocarbon ages and calendar dates

A further limitation on the interpretational value of Lateglacial and early Flandrian ^{14}C measurements and, in particular, how well these quantify rates of climatic and environmental change, stems from the present lack of a reliable method for calibrating the radiocarbon timescale. It is generally accepted that, due to naturally-induced secular variations in atmospheric ^{14}C concentration, the conventional radiometric timescale does not provide a true, or for that matter linear, representation of sidereal time. The effect is well-documented by direct dendrochronological comparisons covering the last eight millennia (Pearson & Stuiver, 1986; Stuiver & Pearson, 1986; Stuiver & Becker, 1986), and by comparisons with U-series dates on fossil corals (Bard et al., 1990). It is now widely suspected that similar, and perhaps even more pronounced, variations in the ^{14}C record will have resulted from the marked changes in concentration of atmospheric carbon dioxide that appear to have occurred at the close of the last glacial (Andree et al., 1986; Becker & Kromer, 1986; Neftel et al., 1988). For example, AMS dating of plant macrofossils from Swiss lake deposits (Amman & Lotter, 1989; Zbinden et al., 1989) suggests the presence of two 'plateaux' which record essentially constant ^{14}C enrichments; one between c. 12,800 and 12,600 years BP and the other around 10,000 years BP.

Given that such variations in atmospheric ^{14}C levels have occurred during the Lateglacial and throughout the present interglacial, it seems reasonable to assume that they were universal in frequency and magnitude. Hence, while conventional radiocarbon ages (in years BP) remain directly comparable (i.e. will provide a basis for time-stratigraphic correlation) features such as the above-mentioned 'plateaux' clearly pose a major difficulty in the establishment of 'absolute' ages estimates, especially during the critical period at the end of the last cold stage.

Total carbon content and the $\delta^{13}C$ record

The scientific incentive

As mentioned above, a correspondence was noted between the stable isotope ($\delta^{13}C$) enrichment values determined as a routine part of ^{14}C measurement, and the sequence of climatic/environmental changes inferred from the palynological record in the Llanilid profile. Several researchers (Oana & Deevey, 1960; Nakai, 1972; Stuiver, 1975) have investigated the possible relationship between environmental conditions and natural ^{13}C enrichment values as recorded by organic residues in freshwater sediments. More recently, Hakansson (1985) has reviewed the various factors likely to have contributed to the $\delta^{13}C$ isotopic signal from organic detritus that accumulated during the last glacial/interglacial transition in lakes in southern Sweden. These lake sediment records show that a marked decrease in the $^{13}C/^{12}C$ ratio coincided with the Late Weichselian/early Holocene chronostratigraphic boundary (Mangerud et al., 1974), although curiously this trend was the opposite of that reported in previous publications (Nakai, 1972; Stuiver, 1975). In view of the range of environmental and biogeochemical factors that could be considered to

influence the residual $^{13}C/^{12}C$ content in lake sediments, Hakansson (1985) concluded that the uncritical use of this isotopic ratio as an indicator of climatic or related environmental change could be misleading.

Table 4. Total carbon content (relative organic productivity) and ^{13}C enrichment values from the Llanilid sediment profile.

Sample Depth (cm)	Wt % Carbon	$\delta^{13}C_{PDB}$%
178	48.7	-28.5
176	46.6	-28.5
172	54.5	-29.6
167	53.0	-29.3
162	48.1	-29.4
157	48.7	-29.1
152	49.6	-29.5
147	46.8	-30.3
142	47.9	-30.4
137	48.4	-29.6
133	48.6	-29.3
132	48.8	-28.9
127	32.5	-27.5
126	32.0	-27.3
123	21.7	-27.6
122	10.8	-26.9
117	7.7	-24.6
112	7.2	-25.0
107	7.8	-25.3
102	8.6	-26.0
100	13.3	-26.7
97	17.3	-27.2
92	24.0	-27.8
87	25.8	-28.4
83	23.1	-27.8
82	23.0	-27.9
77	26.6	-27.7
73	32.0	-27.6
72	39.5	-27.9
67	30.4	-27.7
62	39.8	-27.8
57	46.0	-28.5
55	47.3	-28.8
52	48.6	-29.5
47	46.9	-29.0
42	47.6	-29.2
37	47.1	-29.1
32	47.0	-28.8
30	41.2	-28.2
27	30.1	-27.8
22	19.0	-22.8
18	18.2	-25.1
17	16.6	-26.0
12	14.4	-25.8
7	11.0	-24.8
5	8.4	-25.0
2	3.1	-25.4

However, the number of experimentally-determined data from Lateglacial freshwater sediments is relatively sparse and, moreover, the initial results from the Llanilid material appeared to conform very closely to the pattern recorded in the Swedish lake sediments. Hence, a further series of samples was taken from the Llanilid monoliths in order to establish clearly the trend in stable isotope enrichment throughout the Lateglacial and early Flandrian sequence.

Experimental procedures

Samples of *c.* 1 g sediment were extracted at 5 cm intervals throughout the Llanilid profile, and added to those already obtained for ^{14}C dating. The sample material was dried to constant weight in a vacuum oven (at 40°C) and then homogenised by grinding in an agate mortar. Between 10 and 20 mg of the sediment was weighed into a quartz reaction vessel together with an excess of copper oxide and a few mg of silver foil. The reaction tube was pumped to vacuum (≤ 10^{-2}Torr), flame-sealed and introduced into a muffle furnace. The reaction was then cycled through a temperature gradient to 900°C over a twelve hour period to oxidise the organic carbon. The product carbon dioxide was collected and purified by cryogenic trapping followed by vacuum distillation. Yields of carbon dioxide were recorded and expressed as weight percent carbon relative to the raw sediment, and aliquots of the gas were used to determine the ^{13}C enrichment. Results (Table 4) were measured to an overall analytical precision of 0.05‰ and are expressed relative to the international PDB standard in the familiar enrichment (δ) notation:

$$\delta^{13}C_{PDB}‰ = \left[\frac{R\ sample}{R\ standard} - 1 \right] 10^3$$

where '**R**' is the component ^{13}C/^{12}C ratio.

Organic productivity, δ^{13}C and climatic change

The patterns of organic productivity (wt % C content) and δ^{13}C variation in relation to the depositional timescale measured for the 'humin' component of the organic detritus are shown in Figure 5. At first sight, there appears to be a close correlation between organic productivity, ^{13}C enrichment and climatic change as inferred from the palynological/stratigraphic evidence. Lighter (more negative) δ^{13}C values broadly correspond with warmer episodes, while heavier (less negative) values are associated with colder or cooler periods. Within this general pattern, however, there are two particular features that merit further investigation, and which suggest that a simple climatic relationship may not necessarily obtain.

(i)   The sharp (*c.* 400 year duration) pulse of "isotopically heavy" carbon produced just prior to 12,000 years BP. This corresponds to that part of the Llanilid pollen diagram (Figure 2) which shows a marked decline in *Juniperus* values and which has been interpreted as reflecting climatic deterioration (Walker & Harkness, 1990). Indeed, a short-lived climatic deterioration at around 12,000 years BP has been inferred on pollen-stratigraphic grounds from a number of sites throughout the British Isles (*e.g.* Pennington, 1975; Watts, 1985; Walker & Lowe, 1990). However, there is no evidence in the Llanilid profile for a corresponding

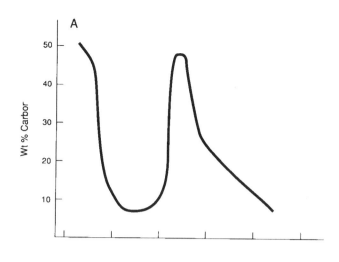

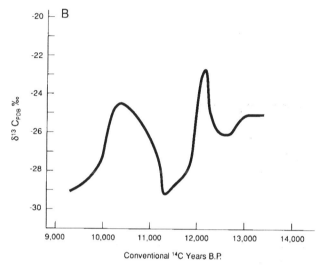

Figure 5 Variations in the stable carbon isotopic composition of the Llanilid sediments as a function of the 'humin' derived ^{14}C timescale. A. Wt % of total carbon in raw (oven dried) sediment. B. δ ^{13}C in total organic carbon.

decline in organic productivity (Figure 4), and hence the ^{13}C enrichment is perhaps more likely to be a reflection of a biogeochemical effect within the former lake basin. In this context, it may be significant that *Chara* oospores, indicative of transient hard-water conditions, are present in this and previous levels in the sediment column (Coope, personal communication).

(ii)   The possible sensitivity of the δ^{13}C signal to climatic change. If a direct relationship between δ^{13}C and ambient temperature is reflected in the data from Llanilid, then the warmest period (lightest δ^{13}C values) during the Lateglacial appears to have occurred just prior to the onset of the Loch Lomond Stadial at *c.* 11,300 years BP. Similarly, the coldest conditions (heaviest δ^{13}C values) within the Loch Lomond Stadial occurred around 1000 years later (at *c.* 10,300 years BP) followed by rapid climatic amelioration at the beginning of the Flandrian. While there is independent biological evidence to support the interpretation of a rapid rise in temperature at the beginning of the present interglacial, and while it is conceivable that the lowest ambient temperatures were achieved late in the Loch Lomond Stadial, there is no support from the fossil record for the contention that the warmest part of the Lateglacial Interstadial occurred around 11,300 years BP. Indeed, the palynological and Coleopteran evidence (Coope, unpublished) point quite unequivocally to the thermal maximum having been reached some 1500 years earlier, a conclusion that is in agreement with data from many other British Lateglacial records (Atkinson *et al.*, 1987). Moreover, if the ^{13}C enrichment curve is a direct reflection of climate, it implies that markedly colder conditions ('heavier' δ^{13}C values) were experienced at *c.* 12,000 years BP than during the Loch Lomond Stadial, an inference which is clearly not borne out by the fossil evidence. It therefore seems apparent that factors other than climate are exerting an influence on the δ^{13}C values in the Llanilid sediment record.

That said, however, the variations in δ^{13}C are intriguing, for long sections of the enrichment curve do seem to parallel climatic trends. Hence, further work on these stable isotopes, in association with related biogeochemical and palaeobotanical analyses, may provide useful additional data in the investigation of patterns of Lateglacial and early Flandrian climatic/ environmental change.

Conclusions

The sequence of Devensian Lateglacial and early Flandrian sediments at Llanilid, South Wales, provided an ideal, almost unique, opportunity to reconstruct a detailed timeframe for palynologically-recorded change in the local and regional environment during this period of fluctuating climate. The ^{14}C dating exercise produced the most secure time-stratigraphic framework so far obtained for Lateglacial events in south-west Britain. However, despite the apparent success of the current research programme, it is clear that radiocarbon dating of the type of organic material found at Llanilid remains beset by difficulties which seriously limit its capabilities in temporal resolution. In essence, the chronological picture from the Llanilid profile still lacks the required depth of focus. This is due partly to lack of knowledge of the taphonomy of the bulk organic detritus and its resulting biogeochemical composition (see Lowe *et al.*, 1988), and partly to the problems that arise from patterns of contemporaneous variations in atmospheric ^{14}C concentration.

The above shortcomings in the present state of applied radiocarbon dating are by no means insurmountable. Indeed, with the development of increasingly sophisticated models to simulate and explain the climatic fluctuations at the last glacial/interglacial transition, there is a clear incentive for research towards improved ^{14}C geochronological resolution. The immediate priorities in this area would seem to fall into three categories:

(i)   The development of a clearer understanding of the biogeochemistry of lake sediments

(ii)   The development of methods for extending the 'high-precision' ^{14}C age calibration data set to cover the past 14,000 years, and perhaps beyond (*e.g.* Bard *et al.*, 1990)

(iii)   An improvement in the analytical precision for routine

AMS measurement of 14C ages greater than *c.* 10,000 years BP.

The collective task clearly demands a collaborative effort with active research input from isotope chronologists, Quaternary research expertise and last, but by no means least, molecular biology.

Acknowledgements

We are grateful to British Coal Opencast for permission to take samples from the opencast working at Llanilid, and to Dr Robert Donelly for drawing our attention to the site and for assistance in the field. Our thanks are also due to Mr B.F. Miller, Mrs E. Wyllie and Mr J. Hannah for analytical support in the NERC Radiocarbon Laboratory at East Kilbride, and to Mr T. Harris who produced the diagrams in the Cartographic Unit at St David's University College, Lampeter. The critical but constructive comments of Dr J.J. Lowe and Dr D.G. Sutherland on earlier drafts of this paper are gratefully acknowledged.

References

AMMAN B & LOTTER A.F. (1989). Late-Glacial radiocarbon- and palynostratigraphy on the Swiss Plateau. *Boreas, 18, 109-126.*

ANDREE M., OESCHGER H., SIEGENTHALER U., RIESEN T., MOELL M., AMMAN B. & TOBOLSKI K. (1986). ^{14}C dating of plant macrofossils in lake sediments. *Radiocarbon, 28,* 411-416.

ATKINSON T.C., BRIFFA K.R. & COOPE G.R. (1987). Seasonal temperatures in Britain during the past 22,000 years, reconstructed using beetle remains. *Nature, 325,* 587-593.

BARD E., HAMELIN B., FAIRBANKS R.G. & ZINDLER A. (1990). Calibration of the ^{14}C timescale over the past 30,000 years using mass spectrometric U-Th ages from Barbados corals. *Nature, 345,* 404-410.

BECKER B. & KROMER B. (1986). Extension of the Holocene dendrochronology by the pre-boreal pine series, 8800-10, 100 BP. *Radiocarbon, 28,* 961-968.

HAKANSSON S. (1985). A review of various factors influencing the stable isotope ratio of organic lake sediments by the change from glacial to post-glacial environmental conditions. *Quaternary Science Reviews, 4,* 135-146.

HEDGES R.E.M. (1991). AMS dating: present status and potential applications. In *Radiocarbon Dating: Recent Applications and Future Potential* (edited by J.J. Lowe), Quaternary Proceedings vol. 1, Quaternary Research Association, Cambridge, pp. 5-10.

LOWE J.J., LOWE S., FOWLER A.J., HEDGES R.E.M. & AUSTIN T.J.F. (1988). Comparison of accelerator and radiometric radiocarbon measurements obtained from Late Devensian Lateglacial lake sediments from Llyn Gwernan, North Wales, UK. *Boreas, 17,* 355-369.

MANGERUD J., ANDERSON S. Th., BERGLUND B.E. & DONNER J.J. (1974). Quaternary stratigraphy of Nordern, a proposal for terminology and classification. *Boreas,* 3, 109-126.

NAKAI N. (1972). Carbon isotopic variations and the palaeoclimate of sediments from Lake Biwa. *Proceedings of the Japan Academy,* 48, 516-521.

NEFTEL A., OESCHGER H., STAFFELBACH T. & STAUFFER B. (1988). CO_2 record in the Byrd ice core 50,000-5000 years BP. *Nature,* 331, 609-611.

OANA S. & DEEVEY E.S. (1960). Carbon 13 in lakewaters and its possible bearing on palaeolimnology. *American Journal of Science,* 258A, 253-272.

PEARSON G.W. & STUIVER M. (1986). High-precision calibration of the radiocarbon timescale, 500 A.D.-2500 B.C. *Radiocarbon,* 28, 839-862.

PENNINGTON W. (1975). A chronostratigraphic comparison of Late-Weichselian and Late-Devensian subdivisions, illustrated by two radiocarbon-dated profiles from western Britain. *Boreas,* 4, 157-171.

PILCHER J.R. (1991). Radiocarbon dating for the Quaternary scientist. In *Radiocarbon Dating: Recent Applications and Future Potential* (edited by J.J. Lowe), Quaternary Proceedings vol. 1, Quaternary Research Association, Cambridge, pp. 27-33.

SCOTT E.M., HARKNESS D.D., COOK G.T., AITCHISON T.C. & BAXTER M.S. (1991). Future quality assurance in ^{14}C dating.

In *Radiocarbon dating: Recent Applications and Future Potential* (edited by J.J. Lowe), Quaternary Proceedings vol. 1, Quaternary Research Association, Cambridge, pp. 1-4.

STUIVER M. & BECKER B. (1986). High-precision decadal calibration of the radiocarbon timescale, A.D. 1950-500 B.C. *Radiocarbon,* 28, 863-910.

STUIVER M. & PEARSON G.W. (1986). High-precision calibration of the radiocarbon timescale, A.D. 1950-500 B.C. *Radiocarbon,* 28, 805-838.

WALKER M.J.C. & HARKNESS D.D. (1990). Radiocarbon dating the Devensian Lateglacial in Britain: new evidence from Llanilid, South Wales. *Journal of Quaternary Science,* 5, 135-144.

WALKER M.J.C. & LOWE J.J. (1990). Reconstructing the environmental history of the last glacial/interglacial transition: evidence from the Isle of Skye, Inner Hebrides, Scotland. *Quaternary Science Reviews,* 9, 15-49.

WATTS W.A. (1985). Quaternary vegetation cycles. In *The Quaternary history of Ireland* (edited by K.J. Edwards & W.P. Warren), Academic Press, London pp 155-185.

ZBINDEN H., ANDREE M., OESCHGER H., AMMAN B., LOTTER A., BONANI G. & WOLFI W. (1989). Atmospheric radiocarbon at the end of the last glacial: an estimate based on AMS radiocarbon dates on terrestrial macrofossils from lake sediments. *Radiocarbon,* 31, 795-804.

Quaternary Proceedings No. 1, 1991 45-53
© Quaternary Research Association, Cambridge

Accelerator and Radiometric Radiocarbon Dates on a Range of Materials from Colluvial Deposits at Holywell Coombe, Folkestone

R.C. Preece

R.C. Preece, 1991 Accelerator and Radiometric Radiocarbon Dates on a Range of Materials from Colluvial Deposits at Holywell Coombe, Folkestone, In *Radiocarbon Dating: Recent Applications and Future Potential* (ed. J.J. Lowe). Quaternary Proceedings No. 1, John Wiley & Sons Ltd, Chichester, pp. 45-53.

Abstract

This paper reports preliminary results of a radiocarbon dating study, involving both conventional and AMS techniques, on a range of materials recovered from colluvial deposits at Holywell Coombe, Folkestone. Materials dated by conventional means include wood, hazel-nuts, organic detritus and tufa. The tufa dates were measured at Harwell and at Gliwice, Poland; all other conventional dates were determined at the Godwin Laboratory (Cambridge). Accelerator dates have been obtained from seeds, charcoal, bone and shells of the land snail *Arianta arbustorum*, and were all measured at the Oxford Laboratory. The results shed light on the following issues : (1) the reproducibility of dates from adjacent sections; (2) comparisons of shell and charcoal dates from the same stratigraphical horizons; (3) comparisons of tufa dates with others based on organic material from the same profiles; (4) assessment of multiple dates from different components of the same material or replicate dates on the same object; and (5) stratigraphical conformity of dates from different laboratories. A replicate biostratigraphical and dating study of two adjacent profiles, reassuringly, gave comparable results. Shell and charcoal samples from a Lateglacial palaeosol likewise produced paired dates that were statistically indistiguishable. All the dates based on organic materials, whether conventional or AMS, were stratigraphically consistent. However, the determinations from all the tufa samples from Holywell Coombe gave apparent ages that were anomalously old. This is thought to be due to contamination from the Chalk bedrock. A parallel study of another tufa in Kent, this time on Atherfield Clay, produced dates very close to expected values.

KEYWORDS: shell; charcoal; tufa; wood; bone; colluvium; palaeosol; Lateglacial; Holocene

Department of Zoology, University of Cambridge, Downing Street, Cambridge CB2 3EJ, U.K.

Introduction

Many valleys in southern Britain, and elsewhere, are floored by thick deposits of colluvium that have accumulated by solifluction, downslope creep or other forms of slope process. These colluvial processes operate mainly on devegetated slopes and give rise to crudely stratified sediments that build up by gradual increments to their surface. Such deposits are important for two reasons. First, they can yield important stratigraphical records of the environmental history of particular areas (*e.g.* Kerney, 1963; Kerney, *et al.* 1964). Second, they can yield detailed sequences of artefacts which can be linked to anthropogenic patterns of colluviation (*e.g.* Bell, 1983).

The dating of such colluvial sequences has until recently been a somewhat frustrating experience. Hitherto, it has usually only been possible to date the occasional organic-rich level which may not always coincide with an important biostratigraphic boundary. The advent of radiocarbon dating by accelerator mass spectrometry (AMS) has completely altered this situation and has, for the first time, enabled the selection of samples independent of the occurrence of bulk organic materials. This paper presents some preliminary results obtained from radiocarbon measurements of samples from colluvial sequences in a valley in Kent.

Location and setting of the site

The Chalk escarpment immediately north of Folkestone, Kent, contains a number of impressive coombe valleys. The main site studied here is the valley immediately west of Sugarloaf Hill, known as Holywell Coombe (Fig. 1). An adjacent valley, immediately east of Castle Hill, is often regarded as the eastern part of the same coombe. The valley has been cut into the Cretaceous Gault Clay, above which there is an important sequence of colluvial deposits covering a large part of the Devensian Lateglacial and most of the Holocene. The stratigraphy essentially consists of a tripartite sequence of (1) basal solifluction debris and associated deposits (Lateglacial) consisting largely of re-deposited Gault Clay, Greensand and Chalk, (2) calcareous tufa (early to mid Holocene), and (3) hillwash (post-Neolithic), but there is much lithological variation, particularly within the Lateglacial (Kerney, *et al.* 1980). The importance of these deposits stems from the completeness of the stratigraphic record and from the fact that waterlogging has inhibited oxidation and has led to the preservation of an array of delicate fossils (*e.g.* pollen, seeds and insects) that are seldom found in such contexts (Fig. 2).

Plans connected with the construction of the Channel Tunnel threatened to destroy this important SSSI, and in July

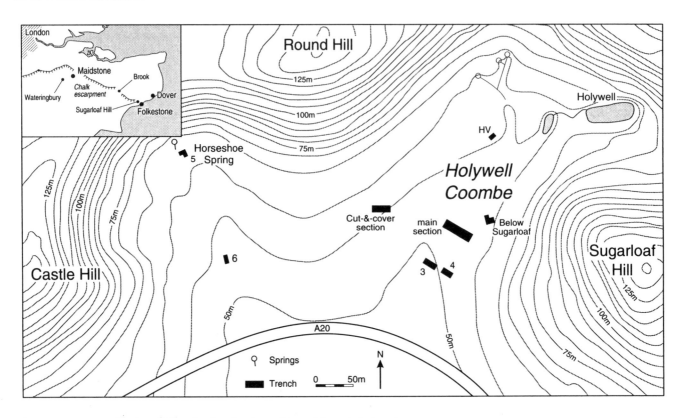

Fig. 1 Location map of Holywell Coombe showing the positions of the sampled sections

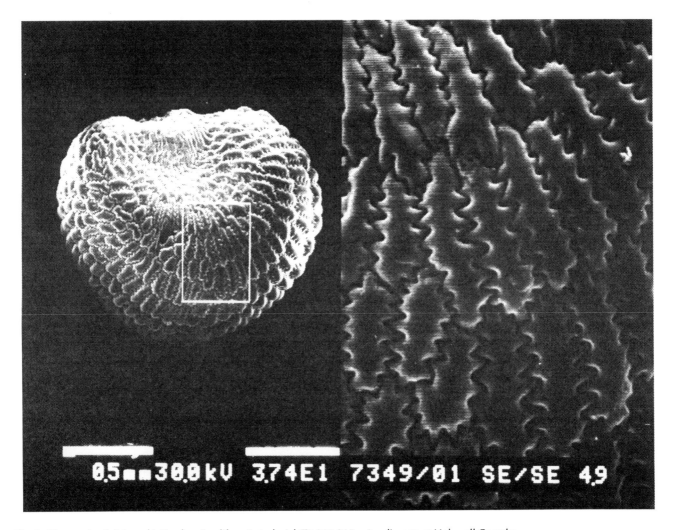

Fig. 2 *Silene vulgaris* (Moench) Garcke. Seed from Lateglacial (T4 310-315cm) sediments at Holywell Coombe.

1987 a major rescue excavation, funded by Eurotunnel, was initiated. This involved a detailed borehole survey, aimed at identifying stratigraphically important areas within the valley and pinpointing the location of organic deposits. A series of deep trenches was then cut in certain critical areas to enable these to be studied in open section and to allow systematic serial sampling (Fig. 3 and 4). A detailed archaeological excavation of the upper hillwash, which was found to yield flint-work and fragments of pottery, was undertaken by the Canterbury Archaeological Trust. The artefacts were ascribable mostly to the early Bronze age, although some Neolithic, Iron Age and later components were also represented. The results of this multidisciplinary project will be published in full elsewhere.

Waterlogging of the basal sediments has resulted in the preservation of a wide variety of datable materials in Holywell Coombe. Previous conventional dates from here have been obtained from wood, hazel-nuts and tufa, and from horizons rich in organic detritus. Further conventional dates have been obtained on these and other materials (e.g. bone) during the present project. In addition, AMS dates have been obtained from seeds, charcoal, bone and shell.

The radiocarbon chronology of several profiles in Holywell Coombe has been established using a combination of both conventional and AMS techniques. In a few instances it has been possible to obtain two, or more, dates from the same stratigraphic horizon, using different datable materials. In other cases, different components of the same material have also been individually dated, and work is in progress to undertake further comparisons of this kind. Five radiocarbon laboratories, Stockholm (St), Cambridge (Q), Harwell (HAR), Gliwice (Gd) and Oxford (OxA) have provided dates from Holywell Coombe. This paper is a preliminary summary of some of these data. Definitive reports will be provided in due course by the laboratories concerned.

Reproducibility of radiocarbon dates from adjacent sections

In most biostratigraphical projects (for example, those dealing with pollen stratigraphy), it is not possible, for reasons of both time and expense, to analyse and date more than one profile. A core is usually taken from the thickest part of a sequence, which is taken to be representative of that sedimentary body. Lake sediments are often favoured because of the quality and resolution of the stratigraphic record they can provide. Recent work has shown that even in such apparently ideal situations, the sediments can be subject to various kinds of disturbance including sediment focussing (Lehman, 1975), coherent slumping (Bennett, 1986) and bioturbation. Moreover, in the instances where multiple profiling has been undertaken, significant differences between central and marginal sequences have been revealed (Davis, Moeller & Ford, 1984).

Colluvial deposits that floor many dry valleys in southern Britain are unlikely to yield such high quality records since there are likely to be many hiatuses within the successions. This handicap can be offset, to some extent, by sampling and analysing multiple profiles. In the present project, no fewer than nine sections, of varying thickness, were studied in Holywell Coombe, and a further two sections immediately west of Castle Hill. Although pollen analyses were undertaken on the basal waterlogged sediments, the upper deposits, above the water-table, were oxidized and found to lack pollen. Land snails, on the other hand, were abundant throughout, and molluscan biostratigraphy was found to be the best means of correlating individual sections.

Table 1 Regional mollusc assemblage zones for the Lateglacial and Holocene of south-east England. The type locality for this zonation scheme is Holywell Coombe, Kent (Kerney et al., 1980). Note that zone y has now been recognized at this site and that the pollen zone equivalents do not always correspond exactly with mollusc zones (see Kerney et al., (1980) for further details).

Mollusc zones (Kent)	Lateglacial & Holocene mollusc assemblage zones (Kent)	Approximate pollen zone equivalents
f	Open ground fauna. As(e), but with appearance of *Helix aspersa*	VIb-VIII
e	Open ground fauna. Decline of shade-demanding species (e.g. *Discus rotundatus*). Re-expansion of *Vallonia*	
d²	Closed woodland fauna. Exansion of *Spermodea, Leiostyla, Acicula*	VIIa
d¹	Closed woodland fauna. Expansion of *Oxychilus cellarius*	
c	Closed woodland fauna. Expansion of *Discus rotundatus*	
b	Open woodland fauna. Expansion of *Carychium* and *Aegopinella*. *Discus ruderatus* characteristic	V
a	As(z), but with decline of bare soil species (notably *Pupilla*) and corresponding expansion of catholic species. Appearance of *Carychium, Vitrea, Aegopinella*	IV
z	Open ground fauna. Restricted periglacial assemblage with *Pupilla, Vallonia, Abida, Trichia*	II-III
y	Open ground fauna. Restricted periglacial assemblage dominated by *Pupilla* and *Vallonia*	I

A detailed outline of the molluscan biostratigraphy of this site has already been published (Kerney et al. 1980). Indeed, so good was the sequence here that it was selected as the type site for a molluscan zonation scheme (Kerney et al. 1980; Table 1). One of the sections (trench 3) created in the present project was located as close as possible to the site of the original stratotype. Reassuringly, there was a close correspondence between both the lithostratigraphy and biostratigraphy of the two profiles, despite the fact that they were studied by different people after an interval of twenty years. Full details of the original dates are given by Welin et al. (1972) and Kerney et al. (1980). Three pairs of dates are now available from these profiles :

	Pit 1 (1969) (Kerney et al., 1980)		Trench 3 (1988)	
Depth (cm)	Material	Lab. ref.	Radiocarbon dates (BP)	Lab. ref.
~ 91-94	organic silt	St-3410	7500 ± 100 7650 ± 80	Q-2716
135-140	wood (+ nuts)	St-3411	8980 ± 100 [8630 ± 120	OxA-2157]
175-180	wood (+ nuts)	St-3395	9305 ± 115 9230 ± 75	Q-2710

Fig. 3 Main section at Holywell Coombe showing the dated 'Allerød soil' (1), the mid-Holocene tufa (2) thinning upslope and the hillwash with two well-developed palaeosols (3&4) (also dated). (Copyright Eurotunnel).

Fig. 4 Deep trench 5 cut near Horseshoe spring. A thick early Holocene tufa sequence overlies some basal Lateglacial deposits. Samples were taken from the narrow gap deliberately left between the trench sheets forming the back-wall. (Copyright Eurotunnel).

Note that whereas the uppermost and lowermost pairs of dates relate to identical stratigraphic horizons, St-3411 and OxA-2157 come from slightly different levels. St-3411 is a conventional date obtained towards the top of a unit of organic tufa *within* mollusc zone b (*sensu* Kerney *et al.* 1980), whereas OxA-2157 is an AMS determination obtained from a higher level (cf 108-112cm) in the equivalent sequence marking the base of mollusc zone c (Fig 5). When this is taken into account, the pairs of dates show remarkable agreement and are statistically indistinguishable at 1σ, despite being measured in three different laboratories.

Comparison of charcoal and shell dates from the same horizon

The shells of land snails have a number of attractions for radiocarbon dating. In southern Britain, for example, they are frequently found *in situ* on sites of archaeological, geological or other interest, where the more standard materials used for dating may be absent. An additional advantage is the fact that stratified molluscan assemblages, and the environmental interpretations that can be drawn from them, have now been intensively studied (*e.g.* Evans, 1972; Kerney, 1977). The principal difficulty encountered in investigating the problems of dating land snails is finding sites where other materials such as charcoal occur within the same stratigraphic horizons. The few direct comparisons of this sort that have been made have had to rely on conventional means of radiocarbon dating (*e.g.* Burleigh & Kerney, 1982). The requirement of only a few milligrams of carbon for AMS dating has reduced the practical problems of obtaining sufficient quantities of datable materials for comparisons of this kind.

In Holywell Coombe, particularly on the flanks of the valley, a well developed palaeosol (the 'Allerød soil') was present within the Lateglacial sequence. This contained a distinctive molluscan fauna dominated by *Trichia hispida*, *Abida secale* and *Arianta arbustorum*, which indicated that it belonged to zone z (cf Table 1). In places, charcoal (mostly *Betula*) was also present in this palaeosol, providing an opportunity to obtain paired dates using both shell and charcoal.

The use of the shells of terrestrial snails for radiocarbon dating has been viewed with much scepticism because the results can be prone to distortion from post-depositional diagenesis or the incorporation of an unknown quantity of inorganic 'dead' carbon from bedrock sources while the snail was alive (Preece, 1980a; Burleigh & Kerney, 1982; Goodfriend & Stipp, 1983; Goodfriend, 1987). The first problem can be largely overcome by careful microscopic examination of individual shells to avoid the selection of those that show evidence of major etching, surface crusts or recrystallization (Yates, 1988). Estimation of the amount of dead carbon incorporated and its effect on producing anomalously old dates is a more difficult problem. Tracer studies have indicated that some species are able to incorporate at least 10-12% of inorganic carbon and in ancient shells this could give rise to over-estimation of age by about 1000 years (Rubin & Taylor, 1963; Rubin, Likins & Berry, 1963). Goodfriend & Stipp (1983) have compared data from snails from limestone and non-limestone areas of Jamaica, and have compared differences between rock-scraping species and those that feed on leaf-litter. They showed that all rock-scraping species and most litter-feeding taxa from limestone areas gave anomalous ^{14}C dates, whereas no anomaly was found in snails from non-limestone areas.

Some species, particularly the non rock-scraping taxa, are therefore more suitable for dating purposes than others, so that a knowledge of the ecology of the species to be dated is desirable.

In this project, the shell of *Arianta arbustorum* was chosen for dating for three reasons. First, it is a litter-feeding species (Grime & Blythe, 1969). Second, it is the largest species in the assemblage, facilitating cleaning during pre-treatment and enabling AMS dates to be based on single shells. Shell fragments of this species are also readily identifiable on account of their characteristic pattern of breakage and diagnostic microsculpture (Preece, 1981). Third, good results (*i.e.* close concordance with charcoal dates) have been obtained with this species in other trials (Yates, 1986, 1988; Yates & Preece, unpublished), suggesting that it does not incorporate significant amounts of 'dead' carbon. The results are given in Table 2.

Table 2. Radiocarbon measurements on various components from the `Allerød soil' at different locations in Holywell Coombe and the neighbouring valley

Main sequence

Lab. ref.	Depth (cm)	Material	δ¹³C (‰)	¹⁴C dates (BP)
OxA-2352	169-172	humic acids	-36.5	11,600 ± 100
OxA-2353	169-172	'reduced carbon'*	-27.6	11,520 ± 90
OxA-2479	169-172	shell (*Arianta*)	-11.1	11,810 ± 120

T10 (Cut-&-cover palaeosol)

Lab. ref.	Depth (cm)	Material	δ¹³C (‰)	¹⁴C dates (BP)
OxA-2089	150-160	charcoal	-26.4	11,370 ± 150
OxA-2158	150-160	shell (*Arianta*)	n.m.	11,500 ± 110

Cherry Garden palaeosol

Lab. ref.	Depth (cm)	Material	δ¹³C (‰)	¹⁴C dates (BP)
OxA-2242	10-20	charcoal	-27.4	11,580 ± 100
OxA-2159	10-20	shell (*Arianta*)	n.m.	11,500 ± 100

Trench HV

Lab. ref.	Depth (cm)	Material	δ¹³C (‰)	¹⁴C date (BP)
OxA-2345	322-327	*Carex/Scirpus* fruits	-25.9	11,530 ± 160

* acid and base insoluble residue (solid).   n.m. = not measured

These are all AMS dates from what is believed to be the same stratigraphic horizon. The paired shell/charcoal dates are in excellent agreement and are statistically identical at 2 σ and strongly support the geological interpretation. The tightness of the clustering is perhaps a little surprising, given that pedogenesis must have taken place over several centuries. Note that unlike all the other dates, which were based on charcoal or shell, OxA-2345 was based on seeds but nevertheless produced a comparable date. In two out of the three pairs of dates, the shells gave very slightly older dates than the charcoal. Burleigh & Kerney (1982) reported a similar finding where the shell dates from a Neolithic soil were accurate to about 5-10%, with a tendency towards over-estimation of age. They predicted that as the main uncertainty is the incorporation of dead carbon by living snails during the formation of their shells, progressively older samples should be proportionally less subject to error. The present study has confirmed this prediction and the discrepancy between the pairs of dates is only about 1-2%. For most geological purposes this is quite acceptable, and the consistency and concordance of these shell dates contributes to an increasing confidence in their use in future studies.

Comparisons of tufa dates with organic dates from the same profiles

Calcareous tufa is a calcium carbonate precipitate deposited by certain springs, streams and seepages. Deposits of tufa are widespread in limestone regions, occurring as discrete bodies locally reaching 10m in thickness and 80 hectares or so in extent. Tufa is variable in texture, occurring as nodules (oncoids), granules and encrustations (e.g. Pazdur et al. 1988; Pedley, 1990). Many tufas are autochthonous and contain good biostratigraphical records covering most of the Holocene (e.g. Kerney et al., 1980), and several have yielded stratified sequences of artefacts (e.g. Preece, 1980b; Evans & Smith, 1983).

In an earlier study (Thorpe et al., 1981), tufa samples from Holywell Coombe were dated (at the Harwell Laboratory) from levels that had already produced dates based on organic materials. This exercise was repeated from a number of other sites in southern Britain. For many sites there was reasonable concordance of dates, but the tufa dates from Holywell Coombe gave apparent ages that were anomalously old by a significant amount.

Similar work has been undertaken in Poland (Pazdur & Pazdur, 1986; 1990; Pazdur et al. 1988; Pazdur, 1988) and elsewhere (e.g. Srdoc et al., 1980). These studies showed a clear correlation between the apparent age (and δ¹³C value) of the tufa and the hydrodynamic conditions of sedimentation. Different tufa lithologies, reflecting contrasting depositional environments, were found to differ in their susceptibility to redeposition and diagenesis. Sinters and 'stromatalites' were regarded as the most reliable materials as their authigenic positions in the profiles are clear and diagenetic changes negligible. Oncoids are also hardly affected by diagenetic changes but their redeposition can be difficult to recognise. Moreover, the innermost part of the oncoid (core) is commonly older than the outer layer (cortex). Moss travertines are believed to present the least reliable material for dating because their massive and porous texture promotes long and relatively free water circulation, leading to dissolution and reprecipitation of carbonate.

Previous studies also indicated that the apparent age of tufa could be influenced by the type of bedrock, a potential source of ¹⁴C-free carbon. To explore this possibility further, additional determinations were undertaken (1) from tufa profiles in Holywell Coombe, where contamination from Chalk bedrock had previously been suspected and (2) from a tufa profile at Wateringbury, a site 50km N.W. of Folkestone, where the bedrock is Atherfield Clay (Kerney et al., 1980). Microscopic examination of washed tufa residues from Holywell Coombe did reveal the occasional Chalk-derived microfossil. To ensure that these were not included in the dated samples, only fractions > 0.5mm were used. An additional check on the 'purity' of the tufa samples to be dated was provided by magnetic susceptibility measurements. Pure authigenic tufa is diamagnetic and will therefore have low frequency susceptibility values which are slightly negative (-0.01 to -0.02 μm³ kg⁻¹). The same would also be true of pure Chalk -derived carbonate, but this often includes a range of other minerals (e.g. marcasite) which give rise to slightly positive susceptibilities (e.g. 0.01 to 0.02 μm³ kg⁻¹). Direct measurement of these can therefore give some indication of the extent of contamination

by detrital carbonate from bedrock sources. It can give no indication of post-depositional diagenetic changes, unless this involved the formation of secondary ferrimagnetic or superparamagnetic minerals.

The results are presented in Table 3. The tufas from both sites studied were deposited in identical spring environments (cf. Kerney et al., 1980) and would be classified essentially as 'sinters', although some small oncoids were included in the basal samples from Wateringbury. The expected ages are based on organic dates relating to the same stratigraphical horizons determined from other profiles in Holywell Coombe. Stratigraphical correlation is based on molluscan biostratigraphy, and the tufa samples chosen for dating were usually taken from the mollusc zone boundaries. The 'error', therefore, does not relate to statistical confidence limits but to the discrepancy between the expected and observed values. In trench HV the estimated age anomaly of the tufa dates varied between 1600 and 6700 years whereas in trench 5 the tufa dates were between 2710 and 3180 years too old. By contrast, the dates from the Wateringbury tufa were all very close to expected values or were only very slightly older (Table 4). It would appear from this limited study that the nature of the bedrock is indeed a very important factor that can seriously distort radiocarbon age determinations from tufa carbonate. It is not clear whether the organic component of the tufa itself would provide more reliable dates, irrespective of bedrock, but it would probably be subject to 'hard-water' error, depending on its exact origin. Further work involving AMS dating of this organic component may resolve this question.

Table 3 Radiocarbon dates from two profiles in Holywell Coombe showing the magnitude of the age anomaly of the tufa determinations. Organic dates from the profiles are also given. The 'expected ages' are based on organic dates relating to the same stratigraphical horizons determined from other profiles in Holywell Coombe (see text for fuller explanation).

Trench HV

Lab. ref.	Depth(cm)	Material	δ¹³C (‰)	¹⁴C dates(BP)	Expected age	'Error'
Gd-5526	90-95	tufa	-9.13	9,200 ± 100	7600	1600
Gd-5521	200-205	tufa	-8.28	12,400 ± 130	8600	3800
Gd-5523	230-235	tufa	-4.88	15,900 ± 180	9200	6700
OxA-2345	322-327	seeds	-25.9	11,530 ± 160	-	
Gd-5527	327-335	tufa	-4.38	17,540 ± 210	11,600	5940
OxA-1752	420-430	bone	-21*	12,280 ± 140	-	

* assumed value

Trench 5

Lab. ref.	Depth(cm)	Material	δ¹³C (‰)	¹⁴C dates(BP)	Expected age	'Error'
Gd-5532	55-60	tufa	-9.05	10,740 ± 110	8000	2740
Gd-5528	155-160	tufa	-8.33	12,230 ± 120	8600	3600
OxA-2088	250-255	charcoal	-26.7	9,460 ± 140	-	
Gd-5533	270-275	tufa	-8.51	12,210 ± 120	9500	2710
Gd-5529	330-340	tufa	-8.62	12,880 ± 140	9700	3180
Q-2721	340-345	organic	-26.58	9,760 ± 100	-	

Multiple dates from different components of the same material or replicate dates on the same object

There have been a number of detailed studies which have compared radiocarbon determinations derived from either different fractions of the same material or   replicate

Table 4 Radiocarbon dates from the Wateringbury tufa, Kent (cf Kerney *et al.*1980). The site is situated away from the Chalk escarpment where the bedrock is Atherfield Clay. The tufa dates from an internally consistent series and are very close to expected values based on molluscan biostratigraphy.

Lab.ref.	Depth(cm)	Material	$\delta^{13}C$ (‰)	^{14}C dates(BP)
Gd-5539	20-25	tufa	-9.35	6,190 ± 70
Gd-5541	60-65	tufa	-9.70	6,460 ± 70
Gd-5536	120-125	tufa	-7.79	7,690 ± 70
Gd-5538	190-195	tufa	-9.08	8,470 ± 70
Q-1425	255-260	wood	-27.2	8,470 ± 190
Gd-5540	360-365	tufa	-9.08	9,720 ± 70
Gd-5542	405-410	tufa	-11.04	10,330 ± 80

measurements on the same piece of wood or bone. Lowe *et al.* (1988), for example, used AMS techniques to measure the ^{14}C activity of the humic acid, lipid and macrofossil cellulose components, as well as the organic residues after treatment with chlorite and HF/HCl. These were compared with conventional radiocarbon dates on bulk sediment samples from equivalent stratigraphic horizons in a replicate core. These determinations enabled them to assess the likely dating errors that could have arisen from contamination by mineral carbon, recycling of older organic compounds and from modern microbial activity. Various palaeopodzolic and Arctic-alpine Brown soils have also been subjected to intensive radiocarbon dating where the acid-soluble 'fulvic acid' fraction was dated separately from the alkali-soluble and acid-insoluble'humic-acid' components (*e.g.* Matthews, 1985; Matthews & Caseldine, 1987; Walker & Harkness (1990) and Harkness & Walker (1991)). They also used conventional techniques to obtain twelve paired dates on both the alkali soluble (humic) and alkali insoluble (humin) organic fractions of a Lateglacial succession in South Wales. Again, this enabled a critical evaluation of the chronology and the quantification of errors arising from various sources of contamination.

A comparable study could not be undertaken at Holywell Coombe, because the sediments were not uniformly organic. However, some limited data along these lines can be presented here. Three determinations have been obtained from the 'Allerød' palaeosol in the main sequence (Table 2). A date (OxA-2352) based on the humic acid fraction of a charcoal sample, produced a very similar date to one based on the

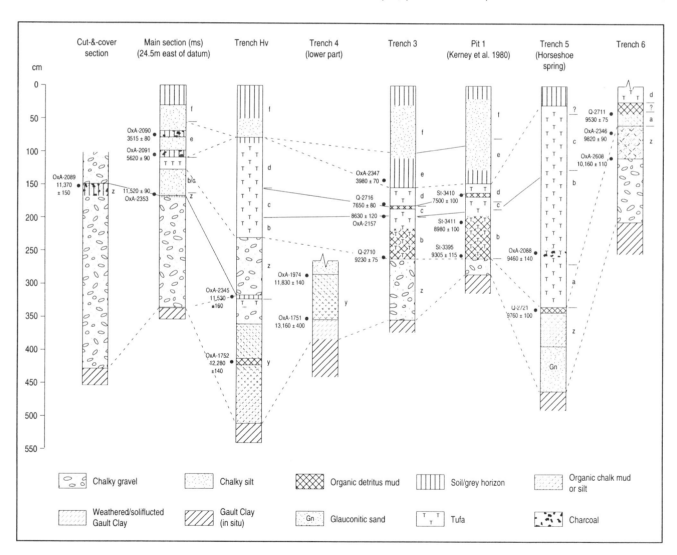

Fig. 5 Stratigraphic context for the organic radiocarbon dates from Holywell Coombe. Simplified logs are given for each of the main profiles. The depths indicated refer mostly to depths below ground level, although the top of the trench 6 profile had been mechanically truncated. Lithological boundaries, which are not necessarily time equivalent, are connected by dashed lines. Biostratigraphic zone boundaries, which are assumed to be isochronous, are joined by solid lines.

'reduced carbon' component (OxA-2353). These two determinations are slightly younger than one based on shell carbonate, although all three measurements are statistically indistinguishable at 2 σ.

Not only have different components of the same material been dated, but replicate determinations on the same object have also proved instructive. Many radiocarbon laboratories have devoted much time to this sort of work and such studies have provided the basis for a number of carefully conducted inter-laboratory comparisons (e.g. Scott et al. 1991). There is scope for obtaining paired AMS/conventional dates on identical wood samples from Holywell Coombe, but as yet this has not been pursued. Replicate AMS dates on a single fragment of wood have been determined:

	Depth (cm)	Material	$\delta^{13}C$ (‰)	^{14}C dates (BP)	Lab. ref.
T6	100-110cm	*Salix* wood	-29.1	$9,900 \pm 100$	OxA-2606
				$10,160 \pm 110$	OxA-2608

Although the means of the two determinations are not identical, the two dates are statistically identical at 2 σ. Not all replicate measurements of the same object have produced such consistent results. Radiocarbon dating, both AMS and conventional, of what are believed to be tusk and teeth from the same Lateglacial mammoth from Condover, Shropshire, produced determinations ranging from $12,300 \pm 180$ to $12,720 \pm 180$ (Lister, 1991), which just fail to overlap at 1 σ from each mean.

Stratigraphical conformity of the radiocarbon dates

Since the organic dates from Holywell Coombe have provided the definitive site chronology against which shell or tufa dates have been compared, it is obviously necessary to evaluate their own internal validity. In the final analysis the appraisal of radiocarbon dates rests on whether they form a stratigraphically consistent series, free from major inversions. Fig. 5 shows the stratigraphical context of most of the organic radiocarbon dates obtained from Holywell Coombe. The dates are shown alongside simplified stratigraphical logs; the mollusc zones are also indicated and provide a valuable biostratigraphic framework. Dashed lines connect lithological boundaries (which are not necessarily time equivalent), whereas solid lines connect biostratigraphic zone boundaries which are assumed to be isochronous, at least within an area the size of Holywell Coombe. The dates based on organic materials obtained during this study were provided 'blind' by the respective laboratories and show remarkable integrity. Not only are they internally consistent but they also accord well with the biostratigraphical data. A full list of the two dozen or so dates is not given here but should be published in the near future.

This study has demonstated the importance of colluvial deposits as a source of palaeoenvironmental and biostratigraphic data. It has high-lighted the necessity of multiple profiling in sequences where hiatuses are likely to occur. It has also shown that by careful selection of species followed by careful pre-treatment, reliable radiocarbon dates can be obtained from the shells of certain land snails. It has also demonstrated the dangers of dating calcareous tufa in areas of Chalk bedrock.

Acknowledgements

I thank the directors of all the radiocarbon laboratories who have provided dates for this study, in particular Dr V.R. Switsur (Godwin Laboratory, Cambridge), Drs A. and M.F. Pazdur (Silesian Technical University, Gliwice, Poland), Dr P.M. Thorpe (formerly at AERE, Harwell), Dr R.E.M. Hedges and Dr R.A. Housley (Oxford). Dr M.J. Sharp (Cambridge) kindly undertook the magnetic susceptibility measurements and provided useful discussion. Richard Burleigh and John Lowe kindly commented on an earlier draft. The work was funded, in part, by Eurotunnel who have continued to show a keen interest in all aspects of the work emanating from the rescue operation.

References

BELL, M. (1983). Valley sediments as evidence of prehistoric land-use on the South Downs. *Proceedings of the Prehistoric Society* 49, 119-150.

BENNETT, K.D. (1986). Coherent slumping of early postglacial lake sediments at Hall Lake, Ontario, Canada. *Boreas*, 15, 209-215.

BURLEIGH, R. & KERNEY, M.P. (1982). Some chronological implications of a fossil molluscan assemblage from a Neolithic site at Brook, Kent, England. *Journal of Archaeological Science* 9, 29-38.

DAVIS, M.B., MOELLER, R.C. & FORD, F. (1984). Sediment focusing and pollen influx. In: Haworth, E.Y. & Lund, J.W.G. (eds) *Lake sediments and environmental history*. Leicester University Press, pp 261-293.

EVANS, J.G. (1972). *Land snails in Archaeology*. Seminar Press, London.

EVANS, J.G. & SMITH, I.F. (1983). Excavations at Cherhill, North Wiltshire, 1967. *Proceedings of the Prehistoric Society* 49, 43-117.

GOODFRIEND, G.A. (1987). Radiocarbon age anomalies in shell carbonate of land snails from semi-arid areas. *Radiocarbon*, 29, 159-167.

GOODFRIEND, G.A. & STIPP, J.J. (1983). Limestone and the problem of radiocarbon dating of land-snail shell carbonate. *Geology*, 11, 575-577.

GRIME, J.P. & BLYTHE, G.M. (1969). An investigation of the relationships between snails and vegetation at the Winnats Pass. *Journal of Ecology*, 57, 45-66.

HARKNESS, D.D. & WALKER, M.J.C. (1991). The Devensian Lateglacial carbon isotope record from Llanilid, South Wales. In *Radiocarbon Dating: Recent Applications and Future Potential* (edited by J.J. Lowe), Quaternary Proceedings Volume 1, Quaternary Research Association, Cambridge, pp. 35-43.

KERNEY, M.P. (1963). Late-glacial deposits on the Chalk of south-east England. *Philosophical Transactions of the Royal Society of London*. B, 246, 203-254.

KERNEY, M.P. (1977). British Quaternary non-marine

Mollusca: a brief review. *In* Shotton, F.W. (ed.) *British Quaternary Studies - Recent Advances.* Oxford University Press, Oxford, pp. 31-42

KERNEY, M.P., BROWN, E.H. & CHANDLER, T.J. (1964). The Late-glacial and Post-glacial history of the Chalk escarpment near Brook, Kent. *Philosophical Transactions of the Royal Society of London* B, 248, 135-204.

KERNEY, M.P., PREECE, R.C. & TURNER, C. (1980). Molluscan and plant biostratigraphy of some Late Devensian and Flandrian deposits in Kent. *Philosophical Transactions of the Royal Society of London* B, 291, 1-43.

LEHMAN, J.T. (1975). Reconstructing the rate of accumulation of lake sediment: the effect of sediment focusing. *Quaternary Research* 5, 541-550.

LISTER, A.M. (1991). Late Glacial mammoths in Britain. *In* Barton, R.N.E., Roberts, A.J. & Roe, D.A. (eds). The Late Glacial in north-west Europe: human adaptation and environmental change at the end of the Pleistocene. *Council for British Archaeology*, Research Report No 77, pp. 51-59.

LOWE, J.J., LOWE, S., FOWLER, A.J., HEDGES, R.E.M. & AUSTIN, T.J.F. (1988). Comparison of accelerator and radiometric radiocarbon measurements obtained from Late Devensian Lateglacial lake sediments from Llyn Gwernan, North Wales, UK. *Boreas,* 17, 355-369.

MATTHEWS, J.A. (1985). Radiocarbon dating of surface and buried soils; principles, problems and prospects. In Richards, K.S., Arnett, R.R. & Ellis, S. (eds) *Geomorphology and soils.* George Allen & Unwin. London & Boston. pp 269-288.

MATTHEWS, J.A. & CASELDINE, C.J. (1987). Arctic-alpine Brown soils as a source of palaeoenvironmental information; further ^{14}C dating palynological evidence from Vestre Memurubreen, Jotunheimen, Norway. *Journal of Quaternary Science,* 2, pp. 59-71.

PAZDUR, A. (1988). The relations between carbon isotope composition and apparent age of freshwater tufaceous sediments. *Radiocarbon,* 30, 7-18.

PAZDUR, A. & PAZDUR, M.F. (1986). [14]C dating of calcareous tufa from different environments. *Radiocarbon,* 28, 534-538.

PAZDUR, A. & PAZDUR, M.F. (1990). Further investigations on ^{14}C dating of calcareous tufa. *Radiocarbon,* 32, 17-22.

PAZDUR, A., PAZDUR, M.F. & SZULC, J. (1988). Radiocarbon dating of Holocene calcareous tufa in southern Poland. *Radiocarbon,* 30, 133-151.

PEDLEY, H.M. (1990). Classification and environmental models of cool freshwater tufas. *Sedimentary Geology,* 68, 143-154.

PREECE, R.C. (1980a). The biostratigraphy and dating of a Postglacial slope deposit at Gore Cliff, near Blackgang, Isle of Wight. *Journal of Archaeological Science* 7, 255-265.

PREECE, R.C. (1980b). The biostratigraphy and dating of the tufa deposit at the Mesolithic site at Blashenwell, Dorset, England. *Journal of Archaeological Science* 7, 345-362.

PREECE, R.C. (1981). The value of shell microsculpture as a guide to the identification of land Mollusca from Quaternary deposits. *Journal of Conchology,* 30, 331-337.

RUBIN, M. & TAYLOR, D.W. (1963). Radiocarbon activity of shells from living clams and snails. *Science,* 141, 637.

RUBIN, M., LIKINS, R.C. & BERRY, E.G. (1963). On the validity of radiocarbon dates from snail shells. *Journal of Geology,* 71, 84-89.

SCOTT, E.M. *et al.* (1991). Future Quality assurance in ^{14}C dating. In *Radiocarbon Dating: Recent Applications and Future Potential* (edited by J.J. Lowe), Quaternary Proceedings Volume 1, Quaternary Research Association, Cambridge, pp. 1-4.

SRDOC, D., OBELIC, B., HORVATINCIC, N. & SLIEPCEVIC, A. (1980). Radiocarbon dating of calcareous tufa: How reliable results can we expect? *Radiocarbon,* 22, 858-862.

THORPE, P.M., HOLYOAK, D.T., PREECE, R.C. & WILLING, M.J. (1981). Validity of corrected ^{14}C dates from calcareous tufa. Actes du Colloque de l'A.G.F. *Formation carbonatées externes, tufs et travertins.* Paris, 9 mai 1981, pp 151-156

WALKER, M.J.C. & HARKNESS, D.D. (1990). Radiocarbon dating the Devensian Lateglacial in Britain: new evidence from Llanilid, South Wales. *Journal of Quaternary Science,* 5, 135-144.

WELIN, E., ENGSTRAND, L. & VACZY, S. (1972). Institute of Geological Sciences radiocarbon dates. III. *Radiocarbon* 14, 331-335.

YATES, T.J.S. (1986). *The selection of non-marine molluscan shells for radiocarbon dating.* Ph.D. thesis, University of London.

YATES, T.J.S. (1988). The detection of diagenetic changes in non-marine shells prior to their submission for ^{14}C dating *In* Olsen, S.L. (ed.) *Scanning electron microscopy in archaeology.* BAR International Series No 452. pp 239-248.

Quaternary Proceedings No. 1, 1991 55-65
© Quaternary Research Association, Cambridge

Radiocarbon Dates from the Antarctic Peninsula Region – Problems and Potential

Svante Björck, Christian Hjort, Ólafur Ingólfsson and Göran Skog

Svante Björck, Christian Hjort, Ólafur Ingólfsson and Göran Skog, 1991 Radiocarbon Dates from the Antarctic Peninsula Region – Problems and Potential, In *Radiocarbon Dating: Recent Applications and Future Potential* (ed. J.J. Lowe). Quaternary Proceedings No. 1, John Wiley & Sons Ltd, Chichester, pp.55-65.

Abstract

A large set of terrestrial, lacustrine, and marine samples from the Antarctic Peninsula region was analysed with respect to ^{14}C age and $\partial^{13}C$ content. The main conclusion is that interpretations and evaluations of ^{14}C dates on marine and especially lacustrine deposits have to be made with great caution. Possibly reliable dates are mixed with largely erroneous dates. Lacustrine surface sediments may be very old due to erosion by bottom freezing. The true marine reservoir effect in this region seems to be in the order of 1200-1300 years. Terrestrial mosses are regarded as giving the most correct ages, but aquatic mosses in lakes without any hardwater effect also seem to give reliable ages. Reservoir effects caused by old ground water are not likely since the lakes mostly lie in permafrost areas and are mainly fed by surface water. The very varying $\partial^{13}C$ values of the lacustrine sediments show that different environment factors, discussed in the text, have influenced the lakes and their sediments. This also stresses the importance of measuring the $\partial^{13}C$ content in Antarctic lake sediments to obtain correct values for the $^{12}C/^{14}C$ relationship. A tephra chronology which is under preparation will be a valuable tool to estimate and understand dating errors.

KEYWORDS: Antarctic; ^{14}C dates; $\partial^{13}C$ values; radiocarbon contaminations; reservoir effects.

Department of Quaternary Geology, Lund University, Tornavagen 13, S-223 63 Lund, Sweden.

Introduction

The first steps towards a radiocarbon chronology for Holocene glacial and climatic events in the Antarctic Peninsula region run the risk of being characterized by errors and misinterpretations of the earliest obtained and scattered ^{14}C dates. However, if these problems are addressed in a methodical way it will gradually be possible to solve them. This paper is an attempt to show how it is possible both to understand and correct a set of sometimes contradictory Antarctic dates providing there is an awareness that at least some of the dates might be seriously affected by different sources of error.

The 56 dated samples referred to in this paper were collected during the 1987/88 POLARSTERN expedition (ANT VI/VII) by Christian Hjort, Olafur Ingólfsson, Wibjörn Karlén and Rolf Zale, and during the 1988/89 SWEDARP expedition by Svante Björck, Christian Hjort, Olafur Ingólfsson, Kerstin Nordström, Anders Wasell and Rolf Zale. The main purpose of the ^{14}C dates is to establish a reliable Holocene ^{14}C chronology upon which glacial geological, palaeoclimatic, and palaeoenvironmental reconstructions can be based. Before this can be done, however, individual dates have to be critically evaluated.

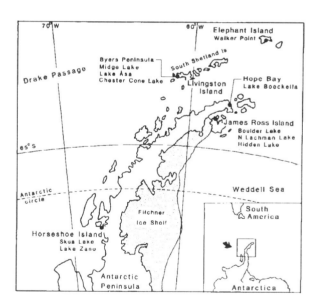

Figure 1 Map of the Antarctic Peninsula region with the sites mentioned in the text.

The study area

Location and climate

The dated samples have been collected from five different sub-areas within the Antarctic Peninsula region (Fig. 1): Elephant Island (61°00'S; 56°00'W), James Ross Island (63°45' - 64°20'S; 57°05' - 58°30'W), Hope Bay (63°24'S; 57°00'W), Byers Peninsula on Livingston Island (64°02'S; 58°42'W) in the South Shetland Archepelago, and Horseshoe Island in Marguerite Bay (67°50'S; 67°20'W). Some of the lake names used, marked with quotation marks in Table 1, are so far unofficial names. Elephant Island, Hope Bay and Livingston Island have a sub-polar maritime climate, with mean

Table 1 Radiocarbon dated samples from Antarctic lakes collected during the 88/89 SWEDARP expeditions. Site names in quotation marks are unofficial, insoluble and soluble means that those fractions have been dated after NaOH treatment of bulk sediment. The dates from Lake Boeckella and Hidden Lake are from Zale and Karlén (1989).

Site	Dated Material	Depth below Sediment Surface (cm)	^{14}C age yrs. BP	δ ^{13}C (‰)	Lab. no.
HOPE BAY					
Lake Boeckella	Sediment	0–2	2275±70	−28.0	St-11990
Lake Boeckella	Sediment	0–23	2585±70	−28.4	St-11746
Lake Boeckella	Sediment	83–113	2140±70	−27.0	St-11747
Lake Boeckella	Sediment	143–173	3720±70	−22.8	St-11620
Lake Boeckella	Sediment	203–233	4085±120	−25.7	St-11748
Lake Boeckella	Sediment	263–293	8615±170	−25.4	St-11619
JAMES ROSS ISLAND					
Hidden Lake	Sediment	70–80	1405±70	−9.1	St-11622
Hidden Lake	Sediment	130–146	3045±70	−12.1	St-11621
"Boulder Lake"	Sediment	12–16	1690±60	−11.1	Lu-3142
"Boulder Lake"	Sediment	16–21	1620±80	−11.4	Lu-3094
"Boulder Lake"	Sediment	21–26.5	1630±100	−12.3	Lu-3095
"N Lachman Lake"	Sediment	35–46	3200±110	−21.8	Lu-3097
LIVINGSTON ISLAND					
Midge Lake	AMS-dates				
	Aquatic Mosses	11–12	755±105		Ua-1216
Midge Lake	Aquatic Mosses	20–21	1340±100		Ua-1217
Midge Lake	Aquatic Mosses	63–56	2635±100		Ua-1218
Midge Lake	Aquatic Mosses	77	2715±100		Ua-1219
Midge Lake	Aquatic Mosses	151	3735±250		Ua-1220
"Chester Cone Lake"	Aquatic Mosses	44–50	2830±130	−25.3	Lu-3170
"Lake Asa"	Sediment	4–7	1140±160	−17.3	Lu-3135
"Lake Asa"	Sediment	17–22	2840±80	−19.1	Lu-3090
"Lake Asa"	Sediment	29–34	2850±80	−18.7	Lu-3128
"Lake Asa"	Insoluble	40–50	7850±150	−23.3	Lu-3134
"Lake Asa"	Soluble	40–50	2240±80	−18.7	Lu-3134a
"Lake Asa"	Sediment	60–65	3110±60	−17.4	Lu-3129
"Lake Asa"	Sediment	83.5–88.5	3570±100	−17.0	Lu-3130
"Lake Asa"	Sediment	91–96	3170±70	−17.0	Lu-3131
"Lake Asa"	Sediment	112–117	3920±90	−18.0	Lu-3132
"Lake Asa"	Sediment	121–126	3930±80	−17.4	Lu-3133
"Lake Asa"	Aquatic Mosses	163–168	3480±140	−25.2	Lu-3089
"Lake Asa"	Soluble	163–168	5240±890	−23.7	Lu-3089a
"Lake Asa"	Aquatic Mosses	205–210	4600±100	−27.8	Lu-3088
"Lake Asa"	Sediment	205–210	5740±180	−24.6	Lu-3088a
HOURSESHOE ISLAND					
"Skua Lake"	Algal flakes	At lake shore	120±45	−8.5	Lu-3100
"Skua Lake"	Sediment	9–13	1440±50	−10.8	Lu-3143
"Skua Lake"	Sediment	14–18	3170±90	−12.1	Lu-3144
"Skua Lake"	Sediment	22–27	3160±60	−11.5	Lu-3099
"Skua Lake"	Sediment	27–30	3440±110	−12.9	Lu-3098
"Lake Zano"	Sediment	2–7	2170±120	−18.5	Lu-3093
"Lake Zano"	Sediment	40–45	2610±100	−15.0	Lu-3136
"Lake Zano"	Sediment	155–160	4580±130	−15.1	Lu-3138
"Lake Zano"	Sediment	230–240	5420±100	−12.1	Lu-3139
"Lake Zano"	Sediment	273–278	5570±120	−13.2	Lu-3148
"Lake Zano"	Sediment	335–340	6510±140	−10.4	Lu-3092
"Lake Zano"	Sediment	378–383	7800±170	−12.9	Lu-3091

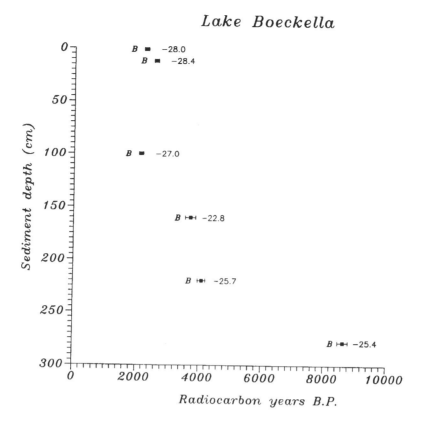

Figure 2 The radiocarbon dates from Lake Boeckella, Hope Bay, related to sediment depth. "B" means that the date was performed on a bulk sediment sample poor in aquatic moss remains. The single standard deviation is marked for each date. The $\delta^{13}C$ values are shown to the right of each sample. The dates are from Zale and Karlén (1989).

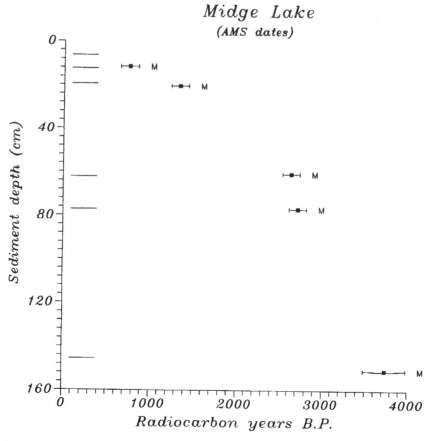

Figure 3 The radiocarbon dates (AMS) from Midge Lake, Livingston Island, related to sediment depth. The lines to the left in the diagram show at what levels tephra was found. "M" means that the date was performed on pure aquatic moss remains. The single standard deviation is marked for each date.

temperatures of –2°C (EI, LI) to –4°C (HB), and are heavily influenced by cyclones moving through Drake Passage. The Equilibrium Line Altitude (ELA) in the maritime Antarctic lies at altitudes between 0 and 150 m a.s.l., depending on topography and exposure. James Ross Island is located in a precipitation shadow behind the mountain range of the Antarctic Peninsula, and the ELA on the northern part of the island probably lies around 400 to 500 m a.s.l. The mean annual temperature there is between –5°C and –6°C. Horseshoe Island in Marguerite Bay is situated between the subpolar glacial regime of southern West Antarctica, with a mean annual temperature of approximately –8°C, but is influenced by cyclonic activity during the austral summer.

Bedrock geology

The bedrock geology varies considerably between the sub-areas, and reflects well the complex geological history of the Antarctic Peninsula region. The South Shetland Islands and Elephant Island are an emerged part of an extension of the submarine Scotia Arc ridge, which links Southern Patagonia with the Antarctic Peninsula. The rocks are composed of sediments, metasediments, volcanics and intrusives, ranging in age from Precambrian to Middle-Late Tertiary. Elephant Island is composed of metamorphic rock of a presumed Cretaceous age. The predominant rocks of Byers Peninsula on Livingston Island are basaltic agglomerates and augite andesites, of Middle Jurassic to Lower Cretaceous age. Sandstones and plant-bearing shales are present in the volcanic sequence at Byers Peninsula. Numerous volcanic plugs penetrate the planated peninsula surface, notably the Chester Cone in the vicinity of Midge Lake and Chester Cone Lake.

The bedrock of northern James Ross Island is composed of Upper Cretaceous marine sediments, which are unconformably overlain by series of Pliocene to Pleistocene volcanics.

The bedrock geology at Hope bay is fairly complex, with components of Upper Paleozoic (Carboniferous to Triassic?) marine sediments, Middle Jurassic to Lower Cretaceous acid tuffs, agglomerates and rhyolite, and Jurassic terrestrial sediments containing numerous fossil plants (Mount Flora).

The bedrock geology of Horseshoe Island is also rather complex. A large part of the island is a plutonic intrusion, belonging to the Andean Intrusive Suite, of presumed Jurassic age. Middle Jurassic to Lower Cretaceous volcanic rocks and metamorphic rocks of uncertain age also occur on the island.

Some factors influencing ^{14}C analysis in Antarctica

The nuclear bomb effect

Since the mid-nineteenth century the atmospheric concentration of ^{14}C has been strongly influenced by anthropogenic effects. Since the onset of the industrial revolution, old CO_2 (free from ^{14}C) has been released to the atmosphere. This dilution of the naturally occuring CO_2 has caused a decrease in the ^{14}C specific activity in the atmosphere; the "Suess effect". On the other hand, the detonation of nuclear bombs in the atmosphere in the early 1960's increased the atmospheric ^{14}C concentration drastically. The maximum ^{14}C level, close to 100% above normal concentration, occured a few years later and reached ca. 60% above normal (Nydal, 1966; Rafter & O'Brien, 1970). Owing to the exchange of CO_2 between the atmosphere and the ocean, a continuous decrease of atmospheric ^{14}C-activity has been observed since the mid

1960's. At present the level has dropped below 20% above pre-bomb normal. Since 1972 the concentration of atmospheric ^{14}C has been the same for the two hemispheres, which shows that the atmosphere is globally mixed within less than 10 years. The latitudinal variations of the "Suess effect" may be considered insignificant, since the rate of ^{14}C dilution has been small.

A clear signal of "bomb ^{14}C" has not been recognised in samples from the Antarctic Peninsula region. In a few ^{14}C determinations of samples from surrounding regions a clear presence of "bomb ^{14}C" has been recognised. Harkness (1979) reported such influence in peat and shells from South Georgia, and Omoto (1972) found the presence of "bomb ^{14}C" in water from a lake in Queen Maud Land. The lack of a clear signal of "bomb ^{14}C" in Antarctic Peninsula samples may be explained by the over-shadowing influence of other reservoir effects (see below).

None of the dates in Table 1 show any clear presence of "bomb ^{14}C", but the low ages of the uppermost dates in Lake Åsa and Skua Lake, compared with the dates below them, might indicate some influence. Not even the dating of the algal (*Phormidium*) flakes (LU-3100), picked on the shore of Skua Lake, show any clear evidence of the bomb effect. These dates will be discussed below, in connection with the lake sediment dates.

The marine reservoir effect

The mean residence time for a ^{14}C atom in the oceans, which contain the main part of all circulating carbon, has been estimated to ca. 1000 years. This means that deep sea water has on the average a radiocarbon age ca. 1000 years older than the contemporaneous atmosphere. Owing to the different mixing rates in the oceans this value varies considerably from ocean to ocean and with depth (e.g. Mangerud, 1972). Water above the thermocline is usually better mixed than deep sea water, and this layer, often referred to as the Mixed Ocean Layer, is modelled as separate from the deep ocean. The surface layer between approximately 40°N and 40°S is considered to be well mixed and has an apparent age between 300 and 500 ^{14}C years (e.g. Figure 2 in Mangerud, 1972). However, at higher latitudes much lower ^{14}C-activities occur, which reflect the greater mixing with deep water, as a consequence of the upwelling of old sea water in these regions.

All reported measurements of ^{14}C-activity of sea surface water around Antarctica have yielded very low values. Rafter (1968), for example, reported a ^{14}C age of 2520 years for surface water collected from the Ross Sea during 1960 and more recent investigations (Williams and Linick, 1974) yielded an apparent age of 760 years as a mean value for surface sea water from the Weddell Sea. Modern marine organisms also show low ^{14}C-activity. For example Curl (1980) reported apparent ages of a recent whale and elephant seal (*Mirounga leonina*) from Livingstone Island as 840 and 970 BP, respectively. Shotton *et al.* (1968) reported a shell date from Potter Cove, King George Island, yielding a ^{14}C age of 850 and 590 BP for the inner and outer fractions, respectively. Other areas in Antarctica show the same pattern with ages for supposed modern samples ranging from ca. 400 to 2000 BP. A freshly killed Weddell Seal (*Leptonychotes weddelli*) in the McMurdo Sound yielded a ^{14}C age of 1385 BP (Noakes *et al.*, 1964), while the remains of the same kind of seal, killed by Scott's Northern Party in North Victoria Land in 1912 (= prebomb time), were dated to 1760 BP (Mabin, 1986). Extreme reservoir effects of up to 5000 yrs have recently been suggested for the East Antarctic shelf seas (Domack *et al.*, 1989).

The four modern samples from James Ross Island and Hope Bay fit well with the above described pattern. The ^{14}C age of 1280 BP for bones of a probable Adelie Penguin *(Pygoscelis adeliae)* from Hope Bay, killed by overwintering members of the Nordenskjold Antarctic expedition in February 1903, is approximately the same as the two datings reported by Stuiver *et al.* (1981) and Mabin (1986) of Emperor Penguins *(Aptenodytes forsteri)* from Victoria Land killed in 1912 with ages of 1310 and 1065, respectively. These samples as well as Mabin's (1986) dating of a Weddell seal (see above), are of special importance since the bomb effect is absent and the industrial effect is negligible. The radiocarbon age of a terrestrial sample from the first decade of this century is *ca.* 70 BP. Thus our best estimate for the reservoir age for marine organisms in the Antarctic Peninsula region is 1200-1300 years.

The ∂^{13}C values

One of the main characteristics of the series of Antarctic lake sediment dates in Table 1 is the generally high ∂^{13}C values, compared with values from lacustrine sediments from lower latitudes. It has been shown that usually each lower-latitude lake has a characteristic ∂^{13}C level, depending on local carbon sources, and that extreme values range at least from –38 ‰ to –7 ‰ (Stuiver, 1975). This range in ^{13}C content corresponds to the range found for 250 terrestrial plants by Troughton (1972). Apart from this natural variation Stuiver (1975) also presented data which clearly suggested that the ^{13}C content decreases with higher latitude. If such a general trend exists, our data series is certainly an anomaly.

Penguin guano gives low ∂^{13}C values (see the Lake Boeckella dates in Table 1 and Fig. 2) and dates on water-mosses also seem to give low ∂^{13}C values (see e.g. Lake Åsa and Chester Cone Lake in Table 1), while the algal flakes (mainly the blue-green algae *Phormidium*) in Skua Lake give high ∂^{13}C values. In spite of the abundance of mosses in Lake Zano's sediments the ^{13}C content there is high. The presence of numerous *Dicranella* species as well as the occurrence of the *Bryum* taxon in Lake Zano's bryoflora might explain these high ∂^{13}C values (see Table 1 and Fig. 5). These are not typical aquatic mosses (Dr R. I. Lewis Smith, British Antarctic Survey; pers. comm.). A test on the aquatic plant *Hygrophila polysperma* made by Park and Epstein (1960) showed that the ∂^{13}C values for this plant, that can grow either terrestrially or under the water, was significantly higher (3.8 to 7.3 ‰) on the underwater sample than on the air grown plant. The reason for this was interpreted to be an effect of the lack of the kinetic step (in the case of the underwater plant) between the atmosphere and the leaf cells. This decreased the isotope fractionation between the plant and the atmospheric CO_2. Park and Epstein (1960) also showed that lichens contain significantly higher ∂^{13}C values (–18 to –21‰) than most terrestrial plants (Wickman, 1952; Craig, 1953), that have a range between *ca.* –23 to –30 ‰. The lack of a vascular system in lichens, which means a different photosynthetic fractionation factor, compared to vascular plants, could account for this difference (Park and Epstein, 1960). Lichens are an important plant group in the catchment areas of the Antarctic lakes. The lichen factor could partly explain the generally rather high ∂^{13}C values that characterize these lakes.

Another factor that increases the sediment's ^{13}C content is anaerobic fermentation in the sediment. This produces CO_2 that is enriched in ^{13}C, as a result of coupled oxidation-reduction reactions. The typically resulting ratio in the sediment is *ca.* –12 ‰ (Deevey *et al.*, 1963), which fits well with the value characterizing at least four of the lakes studied (Hidden Lake and Boulder Lake on James Ross Island, and Skua Lake and Lake Zano on Horseshoe Island). The harsher winter climate at those two islands results in a more prolonged ice cover on the lakes, compared to the lakes in more maritime settings. The ice cover increases the possibility for anaerobic fermentation since the lakes are closed from atmospheric oxygen input for the largest part of the year.

The high ∂^{13}C values in the marine molluscs are expected (Table 2). It is however surprising to find that the only two dates on marine sediments, as identified by their diatom content (the two lowermost dated levels from Skua Lake) do not show significantly higher values than many of the lacustrine sediments. It has been shown by *e.g.* Stuiver (1975) and Håkansson (1985) that changes in the ^{13}C content of lacustrine sediments can sometimes be related to climatic changes, but obviously many other factors can influence the ^{12}C/^{13}C ratio so any single factor is difficult to isolate.

The ^{14}C datings

The moss bank series

Table 2 lists the dated samples from a thick moss bank at Walker Point (*ca.* 200 m above sea level) on Elephant Island (Fig. 1). The moss bank, its evolution and significance will be described by Björck *et al.* in a separate paper. The mosses are completely dominated by *Chorisodontium aciphylum.* According to the uppermost dates the surface of the bank is *ca.* 1600 years old. It has been observed that moss banks on Subantarctic islands may stop growing after having reached a critical height (Dr R. I. Lewis Smith, BAS; pers. comm.). We think that the series of dates from the moss bank show that this type of deposit is probably the best available for constructing a radiocarbon chronology for the past *ca.* 5000 years in the Antarctic Peninsula region for the following reasons: (i) there is a balance between the atmospheric carbon content and the in-take of carbon by the mosses; (ii) old ground-water or old carbon from the bedrock will not influence the carbon content of the mosses; (iii) the absence of any extensive down-growth of roots from the living surface plants excludes recent plant contamination; and (iv) the peat after a while becomes permanently incorporated into the permafrost. An uncontaminated ^{14}C dated peat sequence is the best for dating and to correlate with, for example, lake sediment sequences. Usually the best correlation tools would be pollen stratigraphy or some other terrestrial and regional biostratigraphic method such as insect analysis. However, in Antarctica no such biostratigraphical method is available. In the Antarctic Peninsula region there are other, regionally spread, wind blown particles, namely volcanic ashes or tephra. In a forthcoming paper, Björck, Sandgren and Zale hope to show that by combining a good dating series from the moss bank at Walker Point with a few reliable dates from the lake sediments, all major tephra horizons in the region can be dated to establish a reliable radiocarbon-dated tephra chronology for the Antarctic Peninsula region. Such a chronology has great potential for both estimating dating errors as well as establishing their possible sources.

The lake sediment series

This section will discuss possible sources of contamination in different types of dated material collected from the lake sediments around the Antarctic Peninsula. We will, however,

Table 2 Radiocarbon dated Antarctic marine and terrestrial samples collected during the 1987/88 POLARSTERN and 88/89 SWEDARP expeditions. Note that the Cape Lachman dates were performed on three different mollusc species.

Site	Dated Material	Depth below Sediment Surface (cm)	^{14}C age yrs. BP	$\delta\,^{13}$C (‰)	Lab. no.
ELEPHANT ISLAND					
Walker Point	Terrestrial Mosses	2-5	1680±60	-23.1	Lu-3205
Walker Point	Terrestrial Mosses	17-19	1890±50	-22.2	Lu-2953
Walker Point	Terrestrial Mosses	35-37	2270±50	-22.6	Lu-2994
Walker Point	Terrestrial Mosses	76.5-78.5	3050±50	-22.1	Lu-2993
Walker Point	Terrestrial Mosses	96-98	3210±60	-23.0	Lu-2992
Walker Point	Terrestrial Mosses	121.5-124	3670±60	-22.0	Lu-2991
Walker Point	Terrestrial Mosses	124-125.5	3600±120	-22.0	Lu-2990
Walker Point	Terrestrial Mosses	147.5-149.5	5350±60	-21.5	Lu-2952
JAMES ROSS ISLAND					
Cape Lachman	Lunatia elliptica	On the beach	820±50	+1.1	Lu-3102
Cape Lachman	Yoldia eightsii	On the beach	860±50	+2.4	Lu-3103
Cape Lachman	Nacella concinna	On the beach	920±80	+1.5	Lu-3104
HOPE BAY					
Nordenskiöld exp. refuge	Penguin bones from 1903	Kitchen midden	1280±50	-21.7	Lu-3101

not discuss the reservoir effect (Stuiver & Polach, 1977) in lake water which is caused by old groundwater contaminating the submerged flora and giving the sediments in most lakes an age that is too high (see also Olsson, 1986). The permafrost-surrounded lakes in Antarctica are mostly supplied by surface water from annual snow fields and by direct precipitation, so this effect might be of less significance here than in many other regions. Glaciers, however, may be a serious source of contamination and will be discussed below.

Lake Boeckella

The dates from Lake Boeckella (Table 1 and Fig. 2) and Hidden Lake (Table 1) have already been discussed by Zale and Karlén (1989) and the Lake Boeckella dates are further discussed by Zale (in press). Lake Boeckella (45 m a.s.l.) is the only lake investigated which presently is largely influenced by mammals or birds. The influence comes mainly from a large Adélie Penguin rookery reaching the northwestern shore of the lake. According to Zale and Karlén (1989) the penguins have caused a more or less constant reservoir effect in the lake since it came into being. This conclusion was based on Birkenmajer's (1981) and Barsch and Mäusbacher's (1986) statements that penguins have been present on the Antarctic Peninsula since 5000-6000 BP. Based on the age of the uppermost dates from Lake Boeckella (St-11990 and St-11746 in Table 1), Zale and Karlén estimated the reservoir effect to be 2100 years. However, this led to the strange situation that the dated sample at the 100cm level (St-11747 in Table 1) is also of recent age. We know, however, that the apparent age of the bones of a pre-bomb Adélie penguin are in the range of 1100 (Whitehouse et al., 1987) to 1200 years (Lu-3101 in Table 2), which is probably caused mainly by their main diet of krill *(Euphasia superba)*. Zale (in press) later realized that the estimates made by Zale and Karlén (1989) were too simple and instead tried to apply a

mathematical/chemical model to find a correction factor for each dated level. His main arguments are (i) that the Cu and P enrichments in the sediments can be used as a proxy for the number of penguins at the shore; (ii) that the apparent age of the sediments is a combination of penguin guano (proportional to the penguin proxy) in the sediments and the amount of carbon originating from the local bedrock; and (iii) that the amount of "old" carbon from the bedrock is proportional to the amount of mineral matter in the dated sample. The result of Zale's (in press) model is that penguin guano has at least an apparent age of 1610 years. Compared with Zale and Karlén's (1989) original estimates the changes were not very significant, except for the estimate of the age of the deglaciation sediments, which is revised from *ca.* 8700 BP to 6300 BP.

Although Zale's model (in press) is likely to contain some errors (for example, no account is taken of the inflow of "old" water into the lake from the nearby glacier) it is an interesting attempt to solve some of the problems we encounter when trying to date Antarctic lake sediments: the marine reservoir effect introduced by marine mammals and/or birds and the influence of "old" carbon from the bedrock.

The James Ross Island lakes

Sediments from three different lakes on James Ross Island have been radiocarbon dated (Table 1). Zale and Karlén (1989) published the two datings from Hidden Lake (*ca.* 50 m a.s.l.) and there was no reason to suspect any large errors in those dates. The uppermost dated level consisted of a "peat-like" clay gyttja with *ca.* 25% loss on ignition and the lower dating was made on a clayey algal flake gyttja with 25-30% loss on ignition. The high loss on ignition values is probably the main reason that the dates seem correct. Both samples had high ∂^{13}C values.

High ∂^{13}C values also characterize the samples from

Boulder Lake (ca. 200 m a.s.l.). The dates indicate that the sediments, sandy silty gyttjas with 10-15% loss on ignition, were deposited more or less simultaneously. The derived ages of the samples indicate four possible explanations: (i) extremely low sediment input during the last ca. 1600 years preceded by very high input; (ii) no recent or subrecent sedimentation in this shallow (1.8 m) lake since the lake bottom-freezes in winter which causes erosion of the sediments deposited in the summer; (iii) contamination from the fossil rich Cretaceous bedrock; or (iv) an unknown source of contamination gives the sediment too high ages. We regard alternative two or three as the most likely.

The dating from N Lachman Lake (ca. 25 m a.s.l.) also seems to give erroneous ages (Table 1). The dated sediment consisted of a FeS-coloured gyttja silt with water-mosses. This very shallow lake (0.2 m) most probably undergoes bottom-freezing in winter and most likely also experiences desiccation. The former may prevent new sediments from being deposited and the latter may oxidize and break down the organic material in the sediments. Carbonates in the bedrock could contribute to any contamination since the sediments are poor in organic matter. The date probably therefore only gives a rough estimate of how old the sediments in the lake might be. In contrast to the two other lakes the $\partial^{13}C$ value is low, which might be explained by the high moss content.

The Livingston Island lakes

Midge Lake (ca. 70 m a.s.l.) is the largest lake (ca. 700 x 200 m) on the ice-free Byers Peninsula on Livingston Island. From several corings we know that the lake has a uniform stratigraphy. The dated core was taken at a water depth of 8 m, and the sequence contains, as in most other lakes of the region, a number of tephra layers (Fig. 3). The Midge Lake AMS-radiocarbon dates (Björck et al., 1991) are possibly the most reliable of the radiocarbon dated lake sediments in Table 1, as the accelerator datings were performed on pure moss remains picked out of the sediment. The Midge Lake chronology and the moss bank chronology will form the two main bases for the forthcoming regional tephra chronology.

The primary reason for dating the 44-50 cm level in Chester Cone Lake (ca. 50 m a.s.l.), situated ca. 500 m SSE of Midge Lake, was to date one of the more significant tephra layers on Byers Peninsula. The sediments are very rich in mosses, allowing a conventional dating. We regard the dating as a reliable one since it was obtained from pure mosses. As seems to be the rule with moss-rich or pure moss samples, the $\partial^{13}C$ value is high.

Lake Åsa (35 m a.s.l.) is a shallow (0.9 m), small (ca. 100 x 250 m) lake, situated 1 km WNW of Midge Lake, only about 500 m from the ocean. The radiocarbon dates from Lake Åsa (Fig. 4 and Table 1) give a mixture of "odd", unreliable, and possibly also a few reliable ages. They probably also give us some clues as to the sources of contamination which seem to influence all the bulk sediment dates. The dated material, except for the two moss datings (Lu-3088 and Lu-3089 in Table 1), consists of clay gyttja with a varying amount of moss-remains and loss on ignition values between 5 and 12%, i.e. a type of sediment which in Scandinavia often yields ages that are too old due to redeposited Late Pleistocene (interglacial-interstadial) organic matter (Björck & Håkansson, 1982). Such contamination is somewhat less likely in Antarctica since organic production, even during the interglacials, has been low. The age difference of ca. 5600 years between the NaOH-insoluble and NaOH-soluble fractions on the 40-50 cm level

(Lu-3134 and Lu-3134A in Table 1) suggests that a very serious source of contamination may be involved. The varying ages of the sediment dates (Fig. 4), with seemingly varying errors, indicate that the amount of contaminating material varies between samples. During burning of the samples in the carbon analyser it was noticed that the loss of carbon occurred in two steps (Siv Olsson, Lund; pers. comm.). This indicates that two types of carbon are present in the sediment: one that was formed by organisms living in or around the lake, and a second type that was more difficult to burn. The latter type is possibly carbon from the local bedrock. Fossil bearing shales were observed in slopes east and south of the lake. If this explanation is correct and the relative amount of old carbon/Holocene carbon varies between samples it could explain the large and varying errors.

Unexpectedly the age difference between the moss date and its corresponding (163-168 cm) date of the soluble fraction (Lu-3089 and Lu-3089A) is ca. 600 years larger than the age difference between the other moss date and its corresponding (205-210 cm) bulk sediment date (Lu-3088 and Lu-3088A). This suggests that another contamination source might occasionally be present in the soluble fraction. It should be noted, however that the quoted standard error of Lu-3089A is very large (890 years), so large that the corresponding moss date is well within the double standard deviation, which is not the case with the Lu-3088 and Lu-3088A dates.

The dating of the soluble fraction of the 40-50 cm level (Lu-3134A) does not seem to be erroneous. The errors may be minimal in the uppermost dated level due to the shallow nature of the lake, which renders it sensitive to erosion during bottom-freezing. Preliminary ^{137}Cs and ^{210}Pb measurements of the uppermost 20 cm suggest that no recent sedimentation has occurred in the lake. On the other hand there seems to be some mixing in the uppermost 2 cm between older and recent sediments, which might be attributed to winter ice freezing and melting. Based on the assumptions that (i) the moss dates are fairly accurate, (ii) that the pattern of sedimentation rates is not very different from nearby Midge Lake, (iii) some preliminary tephra correlations are correct, and (iv) that Lu-3135 and Lu-3134A are reasonably accurate, we can roughly estimate the varying errors of the bulk sediment dates in Lake Åsa. If Lu-3135 is excluded the errors seem to vary between ca. 600 and 1200 years. We regard these ages as too old and as mainly an effect of contamination by very old carbon.

The Horseshoe Island lakes

Skua Lake is a small lake (ca. 200 x 140 m) situated just south-west of the BAS hut at Sally Cove. It is at only 3.5 m a.s.l. and its outlet enters innermost Sally Cove. The lake is mainly fed by a stream which enters from the west and drains the higher areas to the southwest. During our visit (March 1989) melting snow drifts around the lake also contributed to the lake water. The vegetation around the lake consists of scattered crustose lichens and the surrounding catchment seems to be a popular nesting ground, mainly for Skuas (Stercorarius skua/maccormickii) but also for penguins. Several corings were carried out and the maximum penetration was 80 cm. On the shores of the lake, flakes of Phormidium (blue-green algae) are found in abundance. The flakes were dated to 120±45 B.P., which is unexpected if they are younger than the onset of the nuclear bomb effect. The uppermost sediments in the cores are dominated by the algal flakes in a sandy, silty algal flake gyttja, 12 to 17 cm thick. Below it in the core, there is a sandy, slightly clayey silt gyttja, often FeS-coloured and with a few shells. We

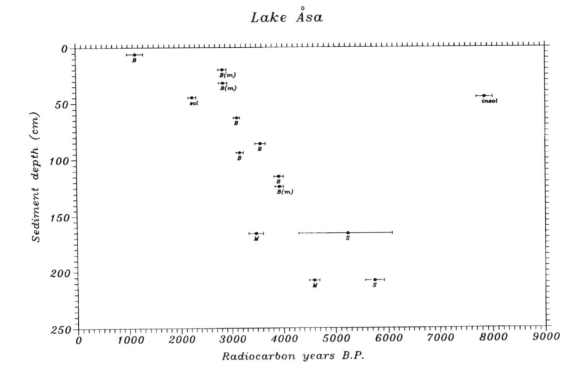

Figure 4 The radiocarbon dates from "Lake Åsa", Livingston Island, related to sediment depth. "B" means that the date was performed on the bulk sediment, "B(m)" on bulk sediment fairly rich in aquatic moss remains, "M" on pure aquatic moss remains, "S" on the sediment without the moss remains, "sol" on the NaOH soluble fraction of the bulk sample. The single standard deviation is marked for each date. The δ^{13}C values are listed in table 1.

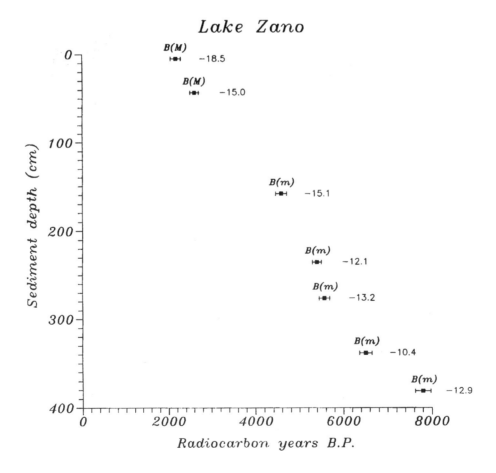

Figure 5 The radiocarbon dates from "Lake Zano", Horeshoe Island, related to sediment depth. "B(M)" means that the date was performed on bulk sediment very rich in aquatic moss remains, "B(m)" on bulk sediment fairly rich in aquatic moss remains. The single standard deviation is marked for each date. The δ^{13}C values are shown to the right of each sample.

interpret this to be a sublittoral marine sediment, which is concurrent with the preliminary diatom analysis (Wasell & Håkansson, in prep.). In the radiocarbon dated core a transitional 14 cm thick layer of sandy silt gyttja, rich in algal flakes, occurs between these two sediment types. The dating series in Table 1 shows a big gap in age at the 13 cm level. This is most probably a marine effect since the 14-18 cm level corresponds to the transitional layer.

The uppermost sediment date is also unexpectedly old. According to the preliminary ^{137}Cs and ^{210}Pb measurements, recent sediments seem to be lacking in Skua Lake. This could also explain the age of the *Phormidium* flakes (Lu-3100). If the lake is too shallow to allow further sedimentation, new sediments washed into the lake, produced in the lake or algae growing on the sediment surface (*cf.* the *Phormidium* flakes), will be frozen into the lake ice during winter and washed ashore or transported out of the lake in spring. The dated *Phormidium* flakes could thus be a mixture of recent, sub-recent and older algae. If the lacustrine sediments are contaminated by "old" carbon, the most likely sources are the marine birds and seals in the vicinity of the lake. The bird guano can also explain the "odd" age of the seemingly fresh *Phormidium* flakes. The lake was probably a sheltered lagoon before it was isolated from the sea. This made it a resting place for seals, which could have increased the marine effect in the three lowermost dated levels (Table 1), and could actually explain why the gap in age is so large between Lu-3143 and Lu-3144.

Lake Zano, (*ca.* 100 m in diameter) is situated about 122 m a.s.l. in front of a large perennial snow field on the southwestern slope of Mt. Searle on Horseshoe Island. Apart from the 4-5 m high ice/snow cliff in the northeastern part of the lake and the outlet in the southwest, the basin is surrounded by relatively steep bedrock slopes. A few crustose lichens were found around the lake and the lake bottom is covered by mosses. The surrounding snow fields deliver meltwater to the lake. The coring was carried out in the western part of the lake, as far away from the large snow field as possible, at a water depth of 5.3 m. One 4.42 m long sediment core was recovered together with three other shorter cores. Altogether seven levels were submitted for ^{14}C analysis from the longest sediment core (Fig. 5 and Table 1). Unfortunately it was not possible to separate the mosses from the sediment fraction in the Lake Zano samples, which means that all dates are on bulk samples, usually rich in mosses.

The pattern of the dated levels suggests that the dates of the whole sequence might be thousands of years too old. Unlike Lake Boeckella or the lakes on Livingston Island there is no reason to ascribe the rather large assumed errors to influence from marine birds or mammals, or to old carbon from the bedrock. It is very likely that the meltwater from the snow field at the northeastern end of the lake produces a hard-water effect through old carbon from the CO_2 in the melting snow/ice. According to Stauffer and Berner (1978) the CO_2 concentrations in polar ice are rather stable, and the content of C in North Greenland ice is 0.05-0.15 g/ton ice (Langway *et al.*, 1965). A steady supply of old carbon from the snowfield will thus function like old groundwater seeping into the lake (Sutherland, 1980), which means that all plants growing in the lake and later becoming parts of the sediment, will be contaminated. The error is probably not constant, since the age difference between the melting ice and the precipitation water must have varied considerably through time. We can conclude that a CO_2 effect from old snow/ice is the most probable source of contamination in Lake Zano. Since such an effect is unlikely

at Skua Lake it implies that these two nearby lakes have been affected by completely different types of contamination sources.

A summary of the problems in dating the lake sediments

Radiocarbon dates on sediments from nine different lakes around the Antarctic Peninsula have clearly shown that they very often seem to yield unexpectedly old ages. Furthermore, the causes for these high ages often seem to be combinations of different contamination sources and processes. These are listed below, together with a summary of the series of bulk sediment dates from lakes (listed in Table 1) that we consider have possibly been influenced by the different sources of error.

1. Contamination by the marine reservoir effect through sea mammals and birds: Lake Boeckella, Skua Lake, Boulder Lake (?), Lake Åsa (?).

2. Contamination from 'old' bedrock carbon: Lake Boeckella, Boulder Lake, N Lachman Lake, Lake Åsa.

3. Contamination from 'old' carbon in the CO_2 from old ice/snow: Lake Zano, Lake Boeckella (?), Boulder Lake (?).

4. Continuous erosion of the surface sediments due to bottom-freezing in winter: Boulder Lake, N Lachman Lake, Lake Åsa, Skua Lake.

5. Oxidation of surface sediments due to periods of desiccation: N Lachman Lake, Boulder Lake (?).

Radiocarbon dates on pure aquatic moss remains seem to yield quite reliable ages. The main reason for this might be that the lake water reservoir effect (Stuiver and Polach, 1977) plays an insignificant role in the permafrost-surrounded Antarctic lakes.

Conclusions

By the very nature of the data presented in this paper, parts of the discussions and interpretations have to be somewhat speculative. We do believe, however, that certain conclusions can be drawn from our experiences on radiocarbon dating in Antarctica.

1. It seems to be more of a rule than an exception that great caution is needed when dates of marine or lacustrine deposits/organisms are interpreted and evaluated. Dates with large errors may be mixed with acceptable dates. High ages on lacustrine surface sediments do not necessarily mean a large reservoir effect or significant contamination, but may also be an effect of the lack of recent sedimentation and/or continuous erosion of lake bottoms. For obvious reasons dates on terrestrial mosses (Table 2) are the most reliable, but our results also suggest that dates on pure aquatic moss remains (Fig. 2) can be trusted.

2. The Antarctic environment does not only produce inconsistent dates, but also results in very variable $\partial^{13}C$ values for lacustrine deposits (Table 1). This emphasises the importance of measuring the $\partial^{13}C$ value to correct for the ^{14}C/^{12}C relationship. In extreme

cases the reported age could be incorrect by hundreds of years without such a correlation.

3. Datings of modern molluscs give ages of 800-900 BP (Table 2), which is in accordance with other studies. These ages are possibly influenced by the "bomb effect", which would suggest that the true marine reservoir effect is more in line with the age of the pre-bomb penguin bones, *i.e.* 1200-1300 BP (Table 2).

4. In order to understand the difficulties and inconsistencies of ^{14}C dates in Antarctica, large sets of dates for different types of material are necessary. The tephra chronology which we are presently preparing for the Antarctic Peninsula region will hopefully be a helpful tool in evaluating ^{14}C dates in this region.

Acknowledgements

We thank the Alfred Wegener Institut für Polar-und Meeresforschung in Bremerhaven and the crew on R/V Polarstern for their logistic support during the 87/88 field season. The Swedish Polar Research Secretariat and the crew on M/S Stena Arctic are acknowledged for all their support during the 88/89 field season. Kerstin Lundahl (Lund) prepared most of the radiocarbon dated samples, Cynan Ellis Evans (BAS, Cambridge) identified the *Phormidium* algae, Ron Lewis Smith (BAS, Cambridge) helped to identify the mosses, Wibjörn Karlén (Stockholm), Kerstin Nordström (Uppsala), Anders Wasell (Stockholm), and Rolf Zale (Umeâ) were our invaluable field work colleagues. To all these people we are very grateful. The Swedish Natural Science Research Council (NFR) financed the radiocarbon datings and the work by SB and ÓI.

References

BARSCH, D. and MÄUSBACHER, R. (1986) New data on the relief development of the South Shetland Islands, Antarctica. *Inter-disciplinary Science Review,* 11 (2), 211-218.

BIRKENMAJER, K. (1981) Raised marine features and glacial history in the vicinity of Arctowski Station, King George Island (South Shetland Islands, West Antarctica). *Bulletin of the Polish Academy of Sciences, Serie Science de la Terra,* 29, 109-117.

BJÖRCK, S. and HÅKANSSON, S. (1982) Radiocarbon dates from Late Weichselian lake sediments in South Sweden as a basis for chronostratigraphic subdivision. *Boreas,* 11, 141-150.

BJÖRCK, S. and HÅKANSSON, S., ZALE, R., KARLÉN, W. & LIEDBERB, B. (1991). A late Holocene lake sediment sequence from Livingston Island, South Shetland Islands, with palaeoclimatic implications. *Antarctic Science* 3, 61-72.

CRAIG, H. (1953) The geochemistry of the stable carbon isotopes. *Geochimica et Cosmochimica Acta,* 3, 53-92.

CURL, J. E. (1980) A glacial history of the South Shetland Islands, Antarctica. *Ohio State University, Institute of Polar Studies,* 63, 1-129.

DEEVEY JR., E. S., STUIVER, M. and NAKAI, N. (1963) Use of light nuclides in Limnology. *In*: Schultz, V. and Klement, A. W. (eds.), *Radioecology Proceedings, 1st National Symposium,* 471-475, Reinhold Publication Corporation, New York.

DOMACK, E. W., JULL, A. J. T., ANDERSON, J. B., LINICK, T. W. and WILLIAMS, C. R. (1989) Application of Tandem Accelerator Mass-spectrometer dating to Late Pleistocene-Holocene sediments of the East Antarctic Continental Shelf. *Quaternary Research,* 31, 277-287.

HÅKANSSON, S. (1985) A review of various factors influencing the stable carbon isotope ratio of organic lake sediments by the change from glacial to post-glacial environmental conditions. *Quaternary Science Reviews,* 4, 135-146.

HARKNESS, D. D. (1979) Radiocarbon dates from Antarctica. *British Antarctic Survey Bulletin,* 47, 43-59.

LANGWAY, C. C. JR., OESCHGER, H., ALDER, B. and RE NAUD, A. (1965) Sampling polar ice for radiocarbon dating. *Nature,* 206, 500-501.

MABIN, M. C. G. (1986) ^{14}C ages for 'heroic era' penguin and seal bones from Inexpressible Island, Terra Nova Bay, North Victoria Land. *New Zealand Antarctic Record,* 7 (2), 19-20.

MANGERUD, J. (1972) Radiocarbon dating of marine shells, including a discussion of apparent age of recent shells from Norway. *Boreas,* 1, 143-172.

NOAKES, J. E., STIPP, J. J. and HOOD, D. W. (1964) Texas A & M University radiocarbon dates I. *Radiocarbon,* 6, 189-193.

NYDAL, R. (1966) Variations in C[14] concentration in the atmosphere during the last several years. *Tellus,* 18, 271-279.

OLSSON, I. U. (1986) Radiocarbon dating. *In*: Berglund, B. E. (ed.) *Handbook of Holocene Palaeoecology and Palaeohydrology,* 275-312, John Wiley & Sons, Chicester.

OMOTO, K. (1972) A preliminary report on modern carbon datings at Syowa Station and its neighborhood, East Antarctica. *Antarctic Records,* 43, 20-24.

PARK, R. and EPSTEIN, S. (1960) Carbon isotope fractionation during photosynthesis. *Geochimica et Cosmochimica Acta,* 21, 110-126.

RAFTER, T. A. (1968) Carbon-14 variations in nature. Part 3: [14]C measurements in the South Pacific and Antarctic Oceans. *New Zealand Journal of Science,* 11, 551-589.

RAFTER, T. A. and O'BRIEN, B. J. (1970) Exchange rates between the atmosphere and the ocean as shown by recent C[14] measurements in the South Pacific. *In*: Olsson, I. U. (Ed.), *Proceedings of the 12th Nobel Symposium on Radiocarbon Variations and Absolute Chronology,* 355-377, Almqvist & Wiksell, Stockholm.

SHOTTON, F. W., BLUNDELL, D. J. and WILLIAMS, R. E. G. (1968) Birmingham University radiocarbon dates II. *Radio carbon,* 10, 200-206.

STAUFFER, B. and BERNER, W. (1978) CO_2 in natural ice. *Journal of Glaciology,* 21 291-299.

STUIVER, M. (1975) Climate versus changes in ^{13}C content of the organic component of lake sediments during the Quaternary. *Quaternary Research,* 5, 251-262.

STUIVER, M. and POLACH, H. A. (1977) Reporting of ^{14}C data. *Radiocarbon,* 19, 355-363.

STUIVER, M., DENTON, G. H., HUGHES, T. J. and FASTOOK, J. L. (1981) History of last glaciation: A working hypothesis. *In*: Denton, G. H. and Hughes, T. J. (eds.), *The Last Great Ice Sheets,* 319-440, John Wiley and Sons, New York.

SUTHERLAND, D. G. (1980) Problems of radiocarbon dating deposits from newly deglaciated terrain: examples from the Scottish Lateglacial. In: Lowe, J. J., Gray, J. M. and Robinson, J. E. (eds.), *Studies in the Lateglacial of North-west Europe,* 139-150, Pergamon Press, Oxford.

TROUGHTON, J. H. (1972) Carbon isotope fractionation by plants. *Proceedings of the 8th International Conference on Radiocarbon Dating, Wellington, New Zealand,* E40-E57.

WHITEHOUSE, I. E., CHINN, T. J. H., VON HOFLE, H. C., McSARENEY, M. J. (1987) Radiocarbon contaminated penguin bones from Terra Nova Bay, Antarctica. *New Zealand Antarctic Record,* 8 (3), 11-23.

WICKMAN, F. E. (1952) Variations in the relative abundance of the carbon isotopes in plants. *Geochimica et Cosmochimica Acta,* 2, 243-254.

WILLIAMS, P. M. and LINICK, T. W. (1974) Cycling of organic carbon in the ocean: Use of naturally occuring radiocarbon as a long and short term tracer. *In: Proceedings of the Symposium on Isotope Ratios as Pollutant Source and Behaviour Indicators,* 153-167, Vienna, 18-22 November 1974.

ZALE, R. (in press) The radiocarbon correction factor in the sediment of Lake Boeckella, Antarctic Peninsula. *GERUM, Department of Geography, University of Umeâ.*

ZALE, R. & KARLÉN, W. (1989) Lake sediment cores from the Antarctic Peninsula and surrounding islands. *Geografiska Annaler,* 71A, 211-220.

Quaternary Proceedings No. 1, 1991 67-73
© Quaternary Research Association, Cambridge

Reconciling the Sea Level Record of the Last Deglaciation with the $\delta^{18}O$ Spectra from Deep Sea Cores

Edouard Bard*, Maurice Arnold, Jean-Claude Duplessy

E Bard, M Arnold, J-C Duplessy, 1991 Reconciling the Sea Level Record of the Last Deglaciation with the δ18O Spectra from Deep Sea Cores, In *Radiocarbon Dating: Recent Applications and Future Potential* (ed. J.J. Lowe). Quaternary Proceedings No. 1, John Wiley & Sons Ltd, Chichester, pp. 67-73.

Abstract

In this paper we use the oxygen isotope record as a transient tracer to study palaeoceanography during the last deglaciation. By using ^{14}C and ^{18}O data obtained on four deep sea sediment cores, we show the presence of a measurable lag between the deglacial $\delta^{18}O$ signal observed in the deep Atlantic and the deep Indo-Pacific oceans. Our study confirms that the major meltwater discharge occured via the North Atlantic and that the thermohaline circulation was operating during the deglacial transition.

KEYWORDS: last deglaciation; transient tracer; thermohaline circulation; oxygen isotopes; accelerator mass spectrometric ^{14}C dating.

Centre des Faibles Radioactivités CNRS-CEA, 91198 Gif-sur-Yvette, France
*also at Lamont-Doherty Geological Observatory of Columbia University, Palisades, NY10964, USA

Introduction

The sea level record of the last deglaciation was recently studied in great detail by Fairbanks (1989). One of the important results of this study is the derivation of the sea level curve to obtain a meltwater discharge history for the period between 18,000 and 6000 ^{14}C years BP (Figure 1). About $50x10^6$ km^3 of isotopically fractionated fresh water was stored in the great glacial ice sheets. Hence the last deglaciation can be viewed as a gigantic tracer experiment. In the case of the Laurentide ice

sheet most of the release of this spiked water occurred via the Mississippi and St Lawrence river systems (Emiliani *et al.* 1975, Broecker *et al.* 1988a) directly to the surface of the Atlantic Ocean. Thereafter the spiked water was redistributed into the world's ocean and the atmosphere at different velocities (Figure 2).

As demonstrated by Fairbanks (1989), sea level history

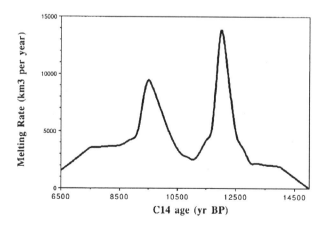

Figure 1 Melting rate during the last deglaciation expressed in km^3 per year. This curve was obtained by calculating the first derivative of the Barbados sea level curve (Fairbanks 1989).

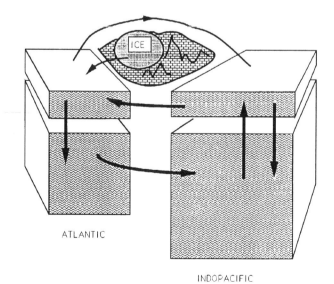

Figure 2 Schematic diagrams of the main pathways for the isotopic signal during the last deglaciation. This diagram is also representing the box-model used for the simulations in Figures 6 and 7.

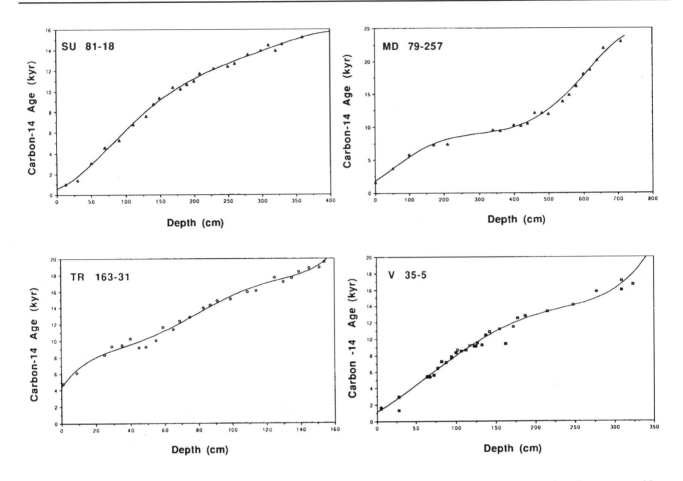

Figure 3 ¹⁴C ages of planktonic foraminifera in the analyzed cores (monospecific samples). The conventional ¹⁴C ages have been corrected for the reservoir age of surface waters, i.e. 400 years for cores SU81-18, MD79-257, V35-5 and 580 years for core TR163-31. For details about the data the reader is referred to Bard *et al.* (1987; SU81-18), Duplessy *et al.* (1990; MD79-257), Broecker *et al.* (1988;V35-5) and Shackleton *et al* (1988; TR163-31).

during the last deglaciation was not at all monotonous but characterized by abrupt changes of melting rates (differences are about one order of magnitude; see Figure 1). Consequently it seems logical that these changes should also be embedded in the δ¹⁸O records measured for the different oceans. However due to oceanic mixing time constants, some differences should be expected if one considers cores from different oceanic reservoirs. In particular the deglacial δ¹⁸O signal is expected to arrive first in the deep Atlantic and then in the deep Pacific. Broecker *et al.* (1988b) attempted to calculate these effects. However the Barbados record was not at their disposal, and a smooth, sigmoidal sea level curve was assumed in their computation.

In order to compare the model calculations with real data it is important to consider the δ¹⁸O records of foraminifera that have been well-dated by accelerator mass spectrometry in order to have comparable ¹⁴C ages with the Barbados discharge record. The transient differences expected during the last deglaciation are on the order of a few hundreds years which is at the limit of the resolution achievable in most deep sea cores. Consequently, it is necessary to work in areas of relatively high sedimentation rates (>10cm/kyr) to minimize the problems of bioturbation.

In this paper we shall first briefly describe the data available to constrain the problem, and then examine how it is possible to interpret them by means of a simple box model of the ocean.

δ¹⁸O signal of the last deglaciation

To a first approximation δ¹⁸O records obtained by analyzing foraminifera in deep-sea sediment reflect the variation of sea water isotopic composition and therefore provide some measure of the sea-level changes caused by glacial cycles (Shackleton, 1967). The oxygen isotope composition of foraminiferal shells, however, depends not only on the isotopic composition of sea water but also on the temperature during foraminiferal growth. δ¹⁸O measurements of benthic foraminifera are thought to provide the best ice-volume proxy because deep-sea temperatures are likely to have been relatively stable during the Pleistocene. However, as demonstrated by Dodge *et al.* (1983), Labeyrie *et al.* (1987) and Shackleton (1987), benthic δ¹⁸O values are not completely free of a temperature dependence, and in particular, the last deglaciation was perhaps accompanied by a warming of about 2 to 3°C in the global ocean deep water. In the following discussion, we shall assume that the temperature imprint on the δ¹⁸O curves does not alter the timing of the records.

For this study we used one core from the North Atlantic, one core from the tropical Indian Ocean and two cores from the Pacific (eastern and western part of the ocean). Core SU81-18 has been raised off Portugal in the Northeastern Atlantic Ocean (Bard *et al.*, 1987a). Core TR163-31 has been collected in the Panama basin and previously studied by Shackleton *et al.* (1988). Core MD79-257 was raised off the

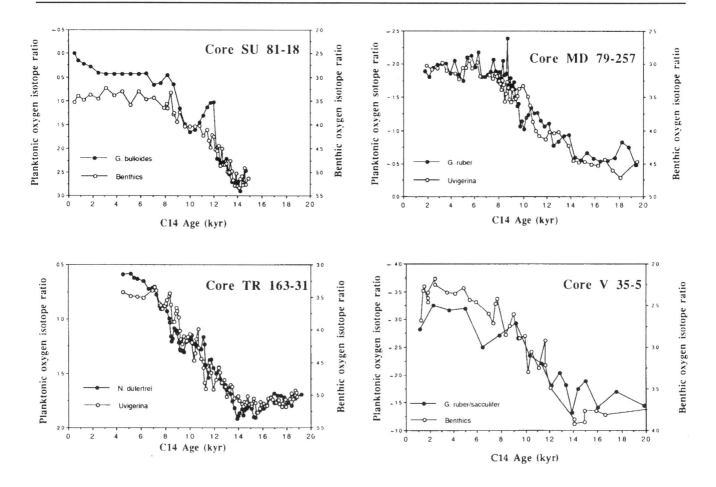

Figure 4 δ¹⁸O records of planktonic and benthic foraminifera plotted against ¹⁴C age (polynomial fits of Figure 3). For details about the data the reader is referred to Bard *et al.* (1987; SU81-18), Duplessy *et al.* (1990; MD79-257, V35-5), Broecker *et al.* (1988; V35-5) and Shackleton *et al.* (1988; TR163-31).

mouth of the Zambezi River in the tropical Indian Ocean (Duplessy *et al.* 1990). Core V35-5 comes from the South China Sea and was primarily studied by Broecker *et al.* (1988c; the planktonic δ¹⁸O data have been produced in Gif-sur-Yvette). In all four cores we have access to AMS ¹⁴C ages on planktonic foraminifera (Figure 3) and δ¹⁸O on both planktonic and benthic foraminifera (Figure 4).

In order to work on a common time scale for the four cores, we first corrected the conventional ¹⁴C ages obtained on monospecific planktonic foraminifera for the reservoir age of surface water at the core location. We have assumed that this correction has not changed through time at the location of the four cores considered for this study. This simplification should not introduce an error larger than a century for middle and low latitude cores (see Bard 1988 for detailed discussion). We then calculated polynomial fits for the ¹⁴C ages versus depth diagrams (Figure 3) to convert depth into time for each core. The δ¹⁸O ratios have been plotted on this timescale in Figure 4. We did not take into account the recent ¹⁴C calibration obtained by mass spectrometric U/Th dating of the Barbados corals (Bard, 1990 a,b,c). Indeed this work shows that during the last deglaciation the ¹⁴C age difference between two samples underestimates the true difference by less than 15% which is negligible for the present study. All ages will thus be reported as conventional ¹⁴C ages in the following discussion.

In order to quantify the time lags between the δ¹⁸O records we used a simple mathematical scheme which calculates the time lag, minimizing the squared area between a particular δ¹⁸O curve and a reference curve (benthic δ¹⁸O of the North Atlantic).

Box-model experiments

A simple box model has been constructed in order to test the various hypotheses concerning the last deglaciation. The model has four oceanic boxes: Surface Atlantic (SA), Deep Atlantic (DA), Surface Indo-Pacific (SIP) and Deep Indo-Pacific (DIP), which have realistic volumes derived from the Pandora model used by Broecker *et al.* (1988b). The main circulation pattern is a thermohaline loop with an additional mixing from the Surface Pacific into the Deep Pacific box. The atmosphere is a simple well mixed box exchanging with the two oceanic surface boxes. This model converges toward realistic radiocarbon content of the deep waters when the residence time of the surface water of the Atlantic is on the order of 40-50 years. This value is equivalent to a flux of North Atlantic Deep Water (NADW) of about 28 Sverdrups, which is in good agreement with the value of 30 Sv determined from the radiocarbon budget of the Atlantic Ocean (Broecker, 1979).

The meltwater input is based on the ¹⁴C ages of reef crest corals from Barbados (Fairbanks, 1989, Figure1). We have assumed that the isotopic composition of the ice sheets was homogeneous. Mix & Ruddiman (1984) predicted a lag between

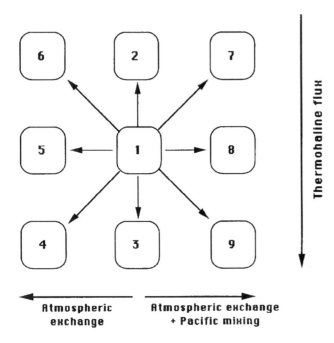

Figure 5 Diagram-caption for Figures 6 and 7. This graph summarizes the different simulations obtained with the box-model. The variable parameters are the thermohaline flux (14 to 84 Sv), the presence of atmospheric isotopic exchange (30% of the input) and Pacific ocean overturn (see text).

the sea level and the mean sea water δ18O signal because the ice caps were not in isotopic equilibrium during the glacial maximum (LGM); the ice melted during the early phase of the last deglaciation being less depleted in 18O than during the end of the climatic transition. However, we decided to neglect this phenomenon because its amplitude is not well constrained and its importance is still strongly debated.

In the first step all the ice sheet is melted into the Atlantic surface and Pacific deep water formation is assumed to be negligible (Figures 5, 6 and 7 cases #1, #2 and #3). The curves obtained for the modern circulation (Figure 6, case #1) show that the DIP and SIP reservoirs lag the DA curve by about 700 years.

Changing the NADW production during the LGM has been tested in cases #2 and #3 of Figure 7. When the modern flux (28 Sv) is reduced by a factor of two (Figure 7 case #2) the calculated lag between DIP and DA is increased by up to 1000 years. For case #3 of Figure 7 we assumed that the conveyor belt was stronger during the deglacial period by a factor of 3. The calculated DIP-DA lag drops to about 400 years.

It is important to note the presence of a strong transient δ18O peak centred at about 12,000 yr BP. This obviously corresponds to the first step of the Barbados sea level curve (Figure 1). However the amplitude of this event depends strongly on the residence time of water in the SA box (0.1 to 0.7 ‰ when the residence time increases by a factor of 6). At first sight, it should thus be easy to track such an event in the deep sea sediment isotopic records. Unfortunately, the δ18O records of planktonic foraminifera usually contain a significant part of temperature imprint and the transient events may be seriously smoothed by

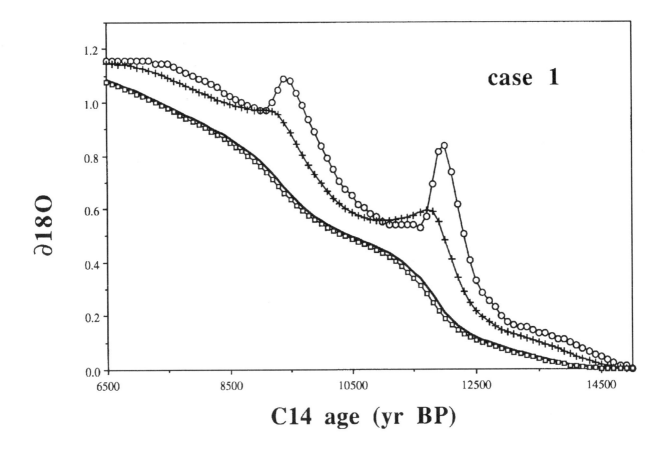

Figure 6 Case 1 of the simulations. This was obtained by assuming a modern thermohaline transport (28 Sv), no atmospheric exchange and no Pacific mixing. The δ18O scale is normalized to a Glacial-Interglacial change of 1.2 ‰. Open circles = Surface Atlantic reservoir (SA), crosses = Deep Atlantic reservoir (DA), thick line = Deep Indo-Pacific reservoir (DIP), open squares = Surface Indo-Pacific reservoir (SIP).

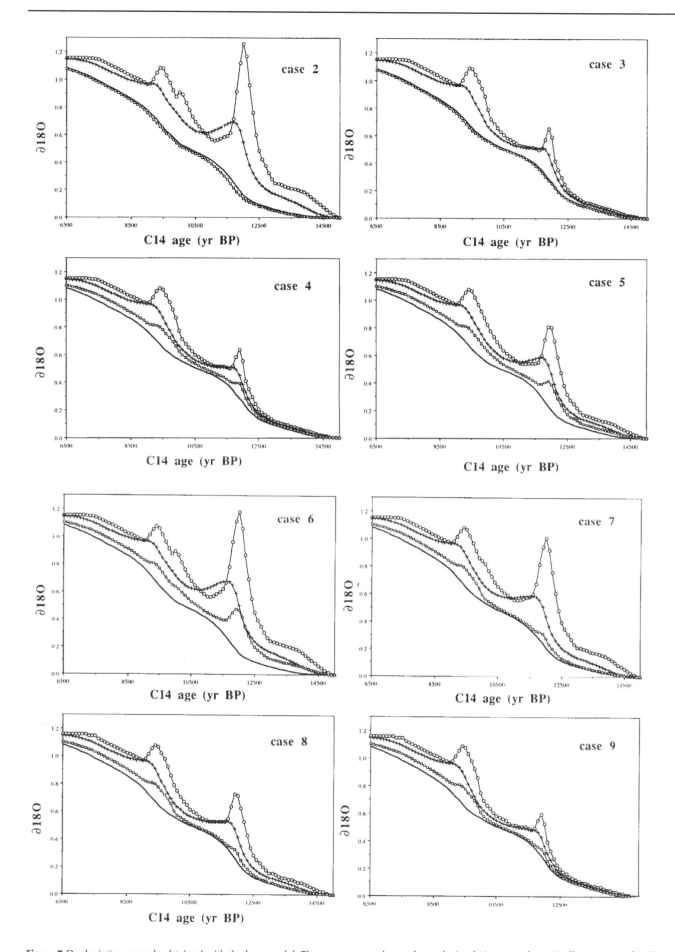

Figure 7 Deglaciation records obtained with the box-model. The parameters relevant for each simulation are schematically represented in Figure 5. Open circles = Surface Atlantic reservoir (SA), crosses Deep Atlantic reservoir (DA), thick line = Deep Indo-Pacific reservoir (DIP), open squares = Surface Indo-Pacific reservoir (SIP).

bioturbation processes (Bard *et al.*, 1987b).

The importance of atmospheric $\delta^{18}O$ exchange has been tested in cases #4, #5 and #6 (Figure 7). The main conclusion is that this process is probably not efficient in changing the lag between the deep reservoirs unless unrealistic values are chosen for the fraction of the $\delta^{18}O$ signal travelling through the atmosphere. In order to illustrate the effect we assumed that 30% of the meltwater isotopic signature is transferred directly into the SIP box. On Figure 7 it can be seen that this mechanism only changes the SIP signal that leads the DIP signal by about 500 years. It should be noted that the value of 30% is highly exaggerated since under modern conditions a value of about a few percent of atmospheric transport can be calculated (Duplessy *et al.*, 1990).

By assuming an important North Pacific mixing we expected to change the DIP signal by redistributing quickly the fraction of the $\delta^{18}O$ signal travelling through the atmosphere. However, due to the mass ratio between the DIP and SIP reservoirs this mechanism proved also to be very inefficient (Figure 7 cases #7. #8 and #9).

Model-data comparison

The best lag estimates of the Deep Indian/Deep Pacific water masses are 800 years for V35-5, 1000 years for core TR163-31 and 800 years for core MD79-257 (Duplessy *et al.*, 1990). These results indicate that the meltwater signal propagated through the whole ocean at a speed which is equal to or slightly lower than expected from the modern circulation (case #1).

The planktonic records of cores TR163-31 and V35-5 lag those of benthic foraminifera by 200 and 800 years, respectively. This suggests that the direct transfer of the deglacial signal through the atmosphere can be considered as negligible when assessing the benthic and planktonic records from the Pacific Ocean.

The planktonic record of core MD79-257 is in phase with the Atlantic benthic record and thus leads the benthic record at that location. Based on the fine structure of the $\delta^{18}O$ record, Duplessy *et al.* (1990) reasoned that a drainage effect from the Zambezi River amplified locally the impact of atmospheric transport, which has no significant relevance for the global ocean.

Conclusions

Based on four deep sea cores raised in three major oceans, we have shown the presence of a measurable lag between the deglacial $\delta^{18}O$ signal observed in the deep Atlantic and the deep Pacific oceans.

Our study confirms that the major meltwater discharge occurred via the North Atlantic which seems reasonable considering the locations of the glacial ice sheets (Denton & Hughes, 1981).

We suggest that the thermohaline circulation was operating during the deglacial transition. The speed of this conveyor belt was similar to or lower than that in today's ocean. There is no clear evidence for a significant isotopic equilibration of the world ocean due to transport of the $\delta^{18}O$ signal through the atmosphere.

References

BARD, E. (1988). Correlation of accelerator mass spectrometry ^{14}C ages measured in planktonic foraminifera: Paleoceanographic implications. *Paleoceanography*, 3, 635-645.

BARD, E., ARNOLD, M., MAURICE, P., DUPRAT, J., MOYES, J. & DUPLESSY, J.C. (1987a). Retreat velocity of the North Atlantic polar front during the last deglaciation determined by accelerator mass spectrometry. *Nature*, 328, 791-794.

BARD, E., ARNOLD, M., DUPRAT, J., MOYES, J. & DUPLESSY, J.C. (1987b). Reconstruction of the last deglaciation: Deconvolved records of $\delta^{18}O$ profiles, micropaleontological variations and accelerator mass spectrometric ^{14}C dating. *Climate Dynamics*, 1, 101-102.

BARD, E., HAMELIN, B., FAIRBANKS, R.G. & ZINDLER, A. (1990a). Calibration of the ^{14}C timescale over the past 30,000 years using mass spectrometric U-Th ages from Barbados corals. *Nature*, 345, 405-410.

BARD, E., HAMELIN, B. & FAIRBANKS, R.G. (1990b). U-Th ages obtained by mass spectrometry in corals from Barbados: sea level during the past 130,000 years. *Nature*, 346, 456-458.

BARD, E., HAMELIN, B., FAIRBANKS, R.G., ZINDLER, A., MATHIEU, G. & ARNOLD, M. (1990c). U/Th and ^{14}C ages of corals from Barbados and their use for calibrating the ^{14}C timescale beyond 9000 years. *Nuclear Instruments and Methods* (in press).

BROECKER, W.S. (1979). A revised estimate for the radiocarbon age of North Atlantic Deep Water. *Journal of Geophysical Research*, 84, 3218-3226.

BROECKER, W.S., OPPO, D., PENG, T.H., CURRY, W., ANDREE, M., WOLFLI, W. & BONANI, G. (1988a). Radiocarbon-based chronology for the $^{18}O/^{16}O$ record for the last deglaciation. *Paleoceanography*, 3, 509-515.

BROECKER, W.S., ANDREE, M., WOLFLI, W., OESCHGER, H., BONANI, G., KENNETT, J. & PETEET, D. (1988b). The chronology of the last deglaciation: implications to the cause of the Younger Dryas event. *Paleoceanography*, 3, 1-19.

BROECKER, W.S., ANDREE, M., KLAS, M., BONANI, G., WOLFLI, W. & OESCHGER, H. (1988c). New evidence from the South China Sea for an abrupt termination of the last glacial period at about 13,500 years ago. *Nature*, 333, 156-158.

DENTON, G.T. & HUGHES, T.J. (1981). *The last great ice sheets*. New York, Wiley Interscience, 484.

DODGE, R.E., FAIRBANKS, R.G., BENNINGER, L.K. & MAURASSE, F. (1983). Pleistocene sea levels from raised coral reefs of Haiti. *Science*, 219, 1423-1425.

DUPLESSY, J.C., BARD, E., ARNOLD, M., SHACKLETON, N.J., DUPRAT, J. & LABEYRIE, L.D. (1990). How fast did the ocean-atmosphere system run during the last deglaciation? Submitted to *Earth Planetary Science Letters*.

FAIRBANKS, R.G. (1989). A 17,000-year glacio-eustatic sea level record: influence of glacial melting rates on the Younger Dryas event and deep-ocean circulation. *Nature*, 342, 637-647.

EMILIANI, C., GARTNER, S., LIDZ, B., ELDRIDGE, K., ELVEY, D.K., HUANG, T.C., STIPP, J.J. & SEVANSON, M.F. (1975). Paleoclimatological analysis of late Quaternary cores from the Northeastern Gulf of Mexico. *Science*, 189, 1083-1087.

LABEYRIE, L.D., DUPLESSY, J.C. & BLANC ,P.L. (1987). Variations in mode of formation and temperature of oceanic deep waters over the past 125,000 years. *Nature*, 327, 477-482.

MIX, A.C. & RUDDIMAN, W.F. (1984). Oxygen-isotope analyses and Pleistocene ice volumes. *Quaternary Research*, 21, 1-20.

SHACKLETON, N.J. (1967). Oxygen isotope analysis and Pleistocene temperature, reassessed. *Nature*, 215, 15-17.

SHACKLETON, N.J. (1987). Oxygen isotopes, ice volume and sea level. *Quaternary Science Reviews*, 6, 183-190.

SHACKLETON, N.J., DUPLESSY, J.C., ARNOLD, M., MAURICE, P., HALL, M. & CARTLIDGE, J. (1988). Radiocarbon age of last glacial Pacific deep water. *Nature*, 335, 708-710.

Quaternary Proceedings No. 1, 1991 75-87
© Quaternary Research Association, Cambridge

Dating of a Maar Lake Sediment Sequence Covering the Last Glacial Cycle

K.M. Creer

K.M. Creer, 1991 Dating of a Maar Lake Sediment Sequence Covering the Last Glacial Cycle, In *Radiocarbon Dating: Recent Applications and Future Potential* (ed. J.J. Lowe). Quaternary Proceedings No. 1, John Wiley & Sons Ltd, Chichester, pp. 75-87.

Abstract

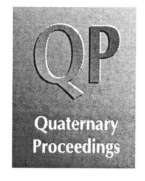

Quaternary Proceedings

A 20m sequence of glacial, late glacial and Holocene sediments deposited in Lac du Bouchet, Haute Loire, France has been subjected to a wide variety of sedimentological, geochemical, palaeontological, palynological and palaeomagnetic studies over the last decade culminating in the European Community funded GEOMAAR project. The results obtained encouraged further work at this site and in 1990, as part of the E.C. funded EUROMAAR project, cores of 50m length were obtained. Hence this site promises to become a standard type section for studies of climatic, environmental and geomagnetic change in western Europe through the last few glacial cycles. This short paper describes the dating control over sediments deposited through the last glacial cycle, using conventional and accelerator radiocarbon ages back to 45 Ka BP and then by correlating warm climatic intervals with the oxygen isotope marine chronology back to 120 Ka BP.

KEYWORDS: Radiocarbon dating; palaeomagnetic record; correlation with oxygen isotope time-scale.

University of Edinburgh, Department of Geology and Geophysics, James Clerk Maxwell Building, King's Buildings, Edinburgh EH9 3JZ

Introduction

Lac du Bouchet is located in the Velay region of the Massif Central, France, at latitude 44.9°N and longitude 3.8°E (Figure 1). The lake, present diameter ~800m, fills a crater of maar origin (Figure 2). The profile of the lake floor slopes steeply from the shore and extends to a flat-bottomed sub-circular area at 27m depth (Figure 3). The rate of sediment accumulation has been remarkably slow compared with that found in most maar lakes and it is most unlikely that there have been appreciable breaks in sediment deposition due to the absence of inflowing and outflowing streams. Though slow for the continental environment, the rate of deposition is nevertheless much faster than in the marine environment. A wide range of multidisciplinary studies has been carried out on the sediments in four phases.

The first studies, made on 6m single piece cores of diameter 6.5 cm taken with pneumatically powered corers (Mackereth, 1958), yielded palynological, sedimentological and palaeomagnetic data extending back to 22 ka BP (de Beaulieu and Reille, 1987; Bonifay and Truze, 1987; Truze, 1988; Thouveny, 1983; Creer *et al.*, 1986; Bonifay *et al.*, 1987).

A second phase, carried out on 9m Mackereth cores, took the palaeomagnetic records back to 35 ka BP (Smith, 1985; Smith and Creer, 1986).

A third phase with 12m cores took the records back to almost 50 ka BP (Reille and de Beaulieu, 1988a; Creer, 1989; Thouveny, 1990; Creer *et al.*, 1991).

The fourth phase was funded by the European Community Science Programme (the GEOMAAR project). Five cores of lengths up to 20m were collected from a raft with a piston corer constructed at the Universities of Kiel and Trier to a Livingstone type design modified by H. Usinger (Figure 4). The cores were taken in sections from 2m to 1m long and their diameter decreased from 8.5 cm through 5 cm to 3 cm from top to bottom of each core. Palynological results are reported in Reille and de Beaulieu (1988b) and palaeomagnetic results in Creer *et al.* (1990); Thouveny *et al.* (1990) and Thouveny *et al.* (1991).

Currently a fifth phase, based on 50m cores taken in September 1990 (the EUROMAAR project), is in progress.

The sediments

The lithology of the sediments has been studied by Truze (1988).

The Holocene is represented by gyttja with some fibrous horizons in the top metre (see Figure 5). The late glacial deposits consist of a clayey silt sediment with intercalated sands and thin silt bands which occur more frequently in *Facies C, E, G, H, I* and *J*. Granular layers comprising variable amounts of clay aggregates are dominant in *Facies D*, which also contains isolated basaltic rock fragments and disseminated authigenic vivianites.

The clay/silt matrix often exhibits a fine stratification with some of the slightly coarser grained laminae consisting of a single grain layer only. Many of the macroscopic sand/silt horizons correspond to multiple turbidites, often with truncated Bouma sequences (Haas, 1989; Haas & Negendank, 1989). The presence of vivianite (iron phosphate) indicates reducing conditions and an absence of sulphide during diagenesis.

Throughout the whole sequence the carbon content varies from 0.5 – 8%. In the absence of significant calcium carbonate, it is a measure of the amount of organic material. Pollen studies, sediment composition and carbon content reveal four

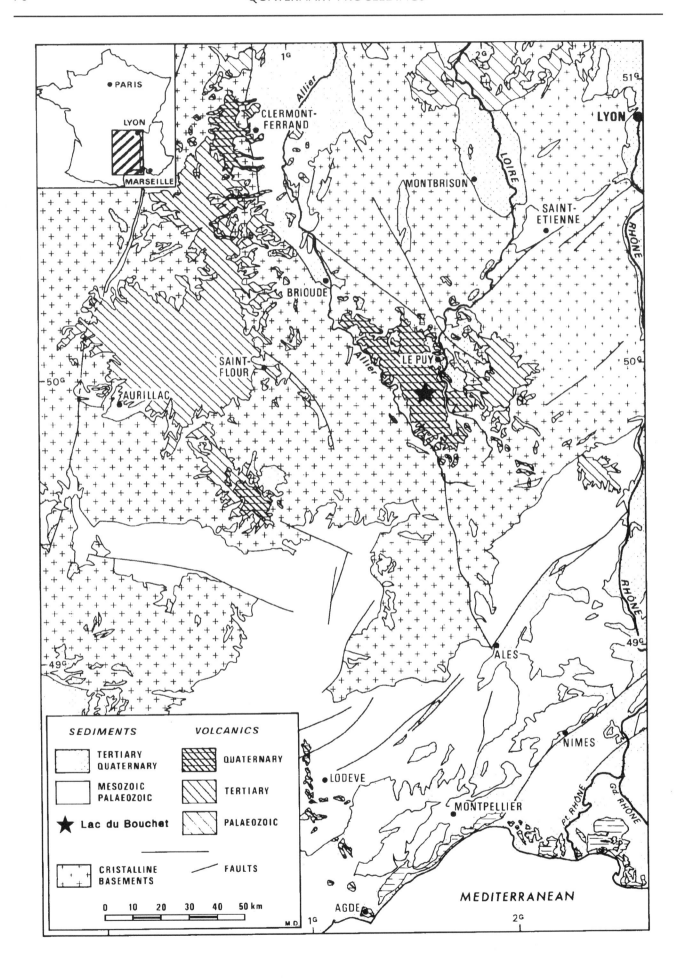

Figure 1 Map showing geographic location of Lac du Bouchet, Haute Loire, France (44.9°N; 3.8°E).

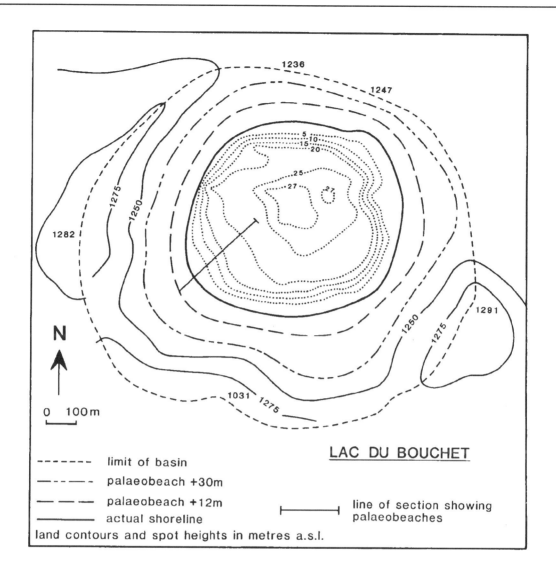

LAC DU BOUCHET

- - - - - - limit of basin
- - · - - · palaeobeach +30m
- - - - - - palaeobeach +12m
———— actual shoreline

├———————┤ line of section showing palaeobeaches

land contours and spot heights in metres a.s.l.

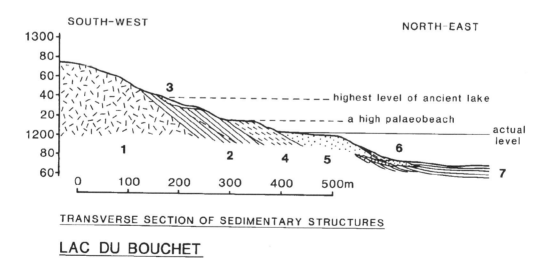

TRANSVERSE SECTION OF SEDIMENTARY STRUCTURES

LAC DU BOUCHET

Figure 2 Sketch map of Lac du Bouchet and the immediately surrounding area. Contours of the lake floor given in metres above mean sea level; contours of lake floor in metres below mean lake level. The transverse section running SW from the middle of the lake shows (1) the volcanic substratum; (2) the highest fossil beach at +30m dating from the end of the middle Pleistocene and (3) deposits accumulated on this beach; (4) the well-defined fossil beach at +12m pre-dating the last interglacial; (5) the present beach and (6) recently deposited gravelly sands and slumps; (7) clayey silts, clays and turbidites of the central part of the lake basin from which cores were taken - after Bonifay and Truze (1987).

3-D MORPHOLOGY OF LAC DU BOUCHET

Contour spacing : 1m.

Figure 3 Three dimensional drawings of the lake basin - contour spacing = 1m - after Hass (1989). For scaling, note that the diameter of the lake is about 800m.

Figure 4 Photograph of the Usinger corer in operation on Lac du Bouchet, September 1989.

warmer climatic phases prior to the Holocene. These are *Facies F* from ~7m to ~9m, *Facies K* at around 16m, *Facies M* between ~18m and ~19m and *Facies O* below ~19.5m depth (see Figure 5).

Magnetic mineralogy

Heavy minerals constitute ~30% by weight of the size fraction 20 – 100µm, less in the finer material, but increasing to ~60% in the coarser sediments.

Magnetite, the dominant magnetic mineral (Smith, 1985), occurs mainly as angular detrital grains which originated from erosion of the volcanic rock forming the crater walls. Diagenetic conditions have precluded growth of magnetic iron sulphides. There is no evidence of bacterial magnetite.

Saturation isothermal remanence, anhysteretic remanence and weak field susceptibility which measure the magnetic mineral content, are all weak in the organic sediments such as *Facies A, K, L* and *O* (Figure 5) and these properties appear to hold the potential of providing a detailed proxy record of past climate, as has been found in some marine sediments (Kent, 1982) and also in some loess deposits(Kukla *et al.*, 1988).

Inter-core correlation

First, the depths of lithological discontinuities observed visually along the different cores were aligned. Photographs of the sectioned cores were also used. However the magnetic susceptibility and palaeomagnetic NRM intensity logs both provide patterns showing more detailed variations than the lithological logs. Therefore the initial correlations were 'fine

tuned' using either susceptibility logs alone (Thouveny *et al.*, 1990), or susceptibility together with NRM intensity (Creer *et al.*, 1990).

Magnetic and palaeomagnetic data from all the individual cores were transferred to a common depth scale prior to stacking. Core D was chosen as 'master' core because, being the longest, it had already been chosen for detailed pollen analytical and sedimentological studies. The depth scales of all three Mackereth core series were transformed to the GEOMAAR core D depth scale.

Radiocarbon dating

A substantial framework of age determinations has been assembled. All the age controls so far obtained are listed in Table 1. Additional information about the radiocarbon age determinations are given in Table 2. Samples for dating were taken from different cores and their downcore depths have been transformed to the core D depth scale. In figure 6a the stratigraphical horizons of the radiocarbon age determinations and palynological age controls for the upper parts of the cores (0 – 10m) are indicated alongside a lithological column drawn up by Pailles (1989).

The 6m Mackereth core series were dated by pollen analytical and conventional radiocarbon measurements on samples taken from core B5. Eight ages back to ~15 ka BP were inferred from pollen studies (de Beaulieu and Reille, 1987; Reille and de Beaulieu, 1988a, b). Five conventional radiocarbon ages (the oldest ~19 ka, BP) were determined at the Gif sur Yvette laboratory. Details of the results are given in Bonifay *et al.* (1987) and Thouveny (1990). Some discrepancy

LAC DU BOUCHET 'GEOMAARS' CORES

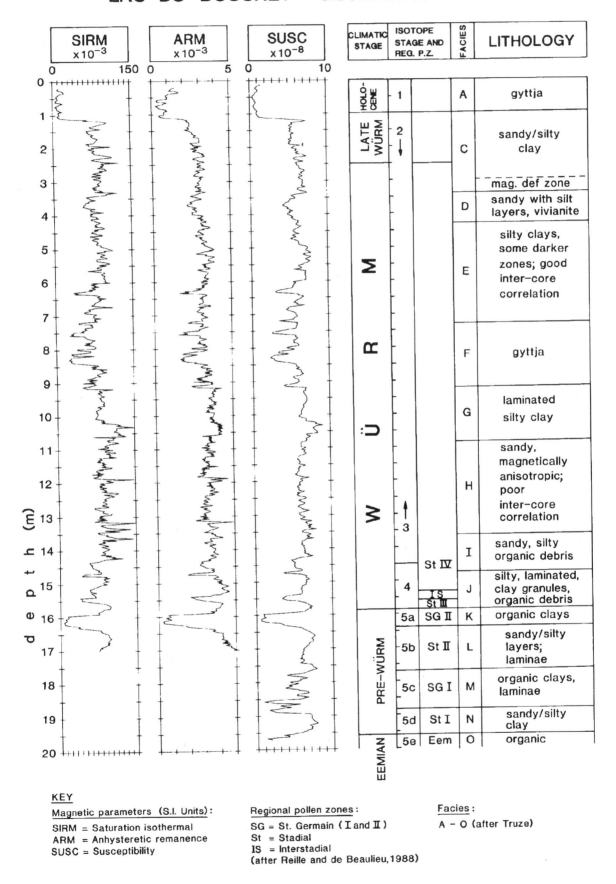

Figure 5 On the left - magnetic parameters (saturation isothermal remanence, anhysteretic remanence and susceptibility) plotted against downcore depth (transformed to core D). On the right - the isotope stages and pollen zones inferred by Reille and de Beaulieu (1988a,b), and descriptions of lithological facies (A - O) (after Truze, 1988).

Table 1 List of age controls

ref no.	core name	depth (cm)*	age (Ka)	source	comments	ref
1	B5	33.2	2.28±0.09	[14]C	conv Gif	1
2	B5	34.7	2.6	pln	base Sub-Atlantic	1
3	B5	54.4	5.5±0.1	[14]C	conv Gif	1
4	B5	58.6	4.7	pln	base Sub Boreal	1
5	B5	78.9	8.0	pln	base Atlantic	1
6	B5	95.6	9.0	pln	base Boreal	1
7	B5	99.0	8.34±0.15	[14]C	conv Gif	1
8	B5	102.0	10.3	pln	base PreBoreal	1
9	B49	106.0	9.70±0.20	[14]C	AMS Oxf	2
10	B5	107.0	10.7	pln	base Dryas 3	1
11	B5	112.0	12.0	pln	Dryas2	1
12	B5	117.0	13.0	pln	base Bolling	1
13	B5	204.5	15.80±0.90	[14]C	conv Gif	1
14	B49	227.0	12.90±0.18	[14]C	AMS Oxf	2
15	F	231.5	17.08±0.20	[14]C	AMS Oxf	8
16	B5	251.0	15.0	pln	base Dryasl	1
17	B49	326.5	14.30±0.25	[14]C	AMS Oxf	2
18	F	338.5	18.16±0.22	[14]C	AMS Oxf	8
19	B49	384.7	19.20±0.30	[14]C	AMS Oxf	2
20	B5	423.0	19.40±1.30	[14]C	conv Gif	1
21	B49	460.7	23.10±0.60	[14]C	AMS Oxf	2
22	F	511.0	27.00±0.60	[14]C	AMS Oxf	8
23	B49	529.4	27.30±0.90	[14]C	AMS Oxf	2
24	B49	590.1	28.50±0.90	[14]C	AMS Oxf	2
25	F	603.0	26.10±0.50	[14]C	AMS Oxf	8
26	D	634.0	22.09±0.46	[14]C	AMS Oxf	8
27	F	710.8	32.90±1.20	[14]C	AMS Oxf	8
28	B49	714.7	30.60±1.30	[14]C	AMS Oxf	2
29	D	753.0	26.81±0.78	[14]C	AMS Oxf	8
30	D	753.0	30.60±1.00	[14]C	AMS Oxf	8
31	B63	804.1	31.08±0.75	[14]C	AMS Tsn	4
32	F	816.3	32.65±1.10	[14]C	AMS Oxf	8
33	D	827.0	>40	[14]C	AMS Oxf	8
34	B63	869.6	40.60±2.40	[14]C	AMS Tsn	4
35	B63	897.9	>39	[14]C	AMS Tsn	4
36	F	918.0	38.20±2.20	[14]C	AMS Oxf	8
37	B63	932.8	38.80±2.20	[14]C	AMS Tsn	4
38	B63	1011.6	>41.5	[14]C	AMS Tsn	4
39	B63	1075	32.20±0.85	[14]C	AMS Tsn	4
40	D	1430	61	pln	δ^{18}O st 3/4a	5
41	D	1590	73	pln	δ^{18}O st 4/5	5
42	D	1925	108	pln	δ^{18}O st 5d	6
43	D	1960	115	pln	δ^{18}O st 5d/e	7

*	:	note that these are core D equivalent depths
ref 1	:	de Beaulieu et al. (1984);
ref 2	:	Smith (1985);
ref 3	:	Creer et al. (1986);
ref 4	:	Creer et al. (1989);
ref 5	:	Hays et al. (1976);
ref 6	:	Shackleton et al. (1983);
ref 7	:	Woillard and Mook (1982);
ref 8	:	Hedges (1989)

Table 2 Radiocarbon ages: laboratory reference numbers

ref no	core name	equ. depth (cm)	age (years bp)	laboratory name	ref nos	%C
01	B5	33.2	2280± 90	Gif	5944	
03	B5	54.4	5500± 100	Gif	5939/40	
07	B5	99.0	8340± 150	Gif	5941	
09	B49	106.0	9700± 200	Ox	OxA-0547 P0990	
13	B5	204.5	15800± 900	Gif	5942	
14	B49	227.0	12900± 180	Ox	OXA-0548 P0991	
15	F	231.5	17080± 200	Ox	OxA-2056 P2483	
17	B49	326.5	14300± 250	Ox	OxA-0549 P0992	
18	F	338.5	18160± 220	Ox	OxA-2057 P2482	
19	B49	384.7	19200± 300	Ox	OxA-0550 P0993	
20	B5	423.0	19400±1300	Gif	5943	
21	B49	460.7	23100± 600	Ox	OxA-0551 P0994	
22	F	511.0	27000± 600	Ox	OxA-2058 P2481	
23	B49	529.4	27300± 900	Ox	OXA-0552 P0995	
24	B49	590.1	28500± 900	Ox	OxA-0553 P0996	
25	F	603.0	26100± 500	Ox	OxA-2059 P2480	
26	D	634.0	22090± 460	Ox	P2474	
27	F	710.8	32900±1200	Ox	OxA-2060 P2479	
28	B49	714.7	30600±1300	Ox	OxA-0554 P0997	
29	D	753.0	26810± 780	Ox	P2475 . 0	
30	D	753.0	30600±1000	Ox	OxA-2154 P2475 .1	
31	B63	804.1	31080± 750	T	AA-1701	0. 67
32	F	816.3	32650±1100	Ox	OxA-2061 P2478	
33	D	827.0	>40000	Ox	OxA-2055 P2476	
34	B63	869.6	40600±2400	T	AA-1702 .	1.17
35	B63	897.9	>39000	T	AA-1703	1. 02
36	F	918 .0	38200±2200	OX	OXA-2062 P2477	
37	B63	932.8	38800±2200	T	AA-1704	0. 93
38	B63	1011.6	>41500	T	AA-1705	1.10
39	B63	1075.0	32200± 850	T	AA-1706	0. 34

1. Tucson: all samples treated same way; carbonates and base soluble organics (humates etc.) removed. Residual material combusted: date is on total organics. #38 - comes from lower 'disturbed' zone - sample possibly from slumped layer.

2. Oxford: P2477-2483 - uniform treatment - extraction with acid, alkali and acid. There was insufficient 'humic' acid to date this fraction alone.
OxA-2 055 (P2 4 7 6) - >4O ka - the ' old ' age gives confidence that no significant background C-14 was added during the combustion stage. # 17 - faulty preparation - could only be dated on one wire.
OxA-2 4 7 4 and 2 4 7 5 - done early on, technical grounds for being unhappy with 2475.

between the radiocarbon ages and those deduced from pollen studies has been noted especially for the Dryas III to Dryas II interval (Creer et al., 1986; Smith and Creer, 1986). Problems in dating this interval are not restricted to Lac du Bouchet and are comprehensively discussed in Lowe et al. (1988).

Twenty-three accelerator ages were obtained. Eight were made on core B49 from the 9m Mackereth series with the High Energy Mass Spectrometer (HEMS) at Oxford and the results are reported in Smith (1985) and in Smith and Creer (1986).

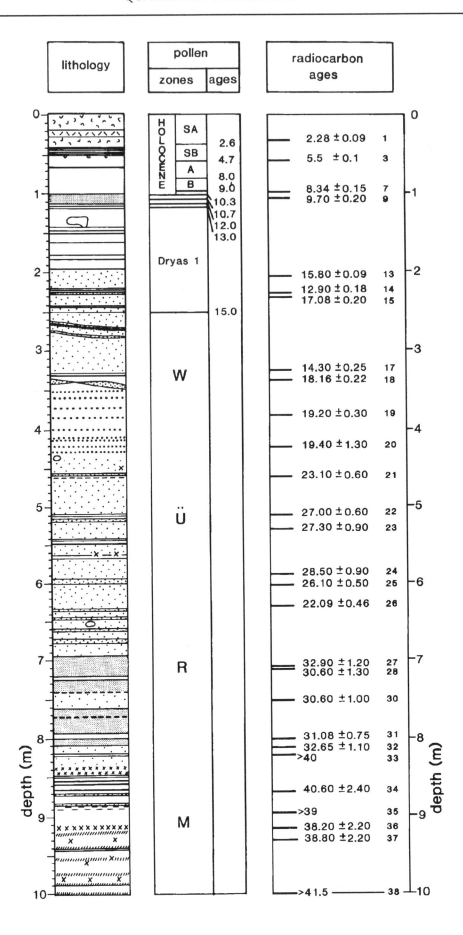

Figure 6a A lithological column (after Pailles, 1989) is shown on the left, the middle columns show climatic zones identified from pollen studies by Reille and de Beaulieu (1988a) and the right hand column indicates the stratigraphic horizons of the radiocarbon age determinations and reference numbers to their entries in Table 1.

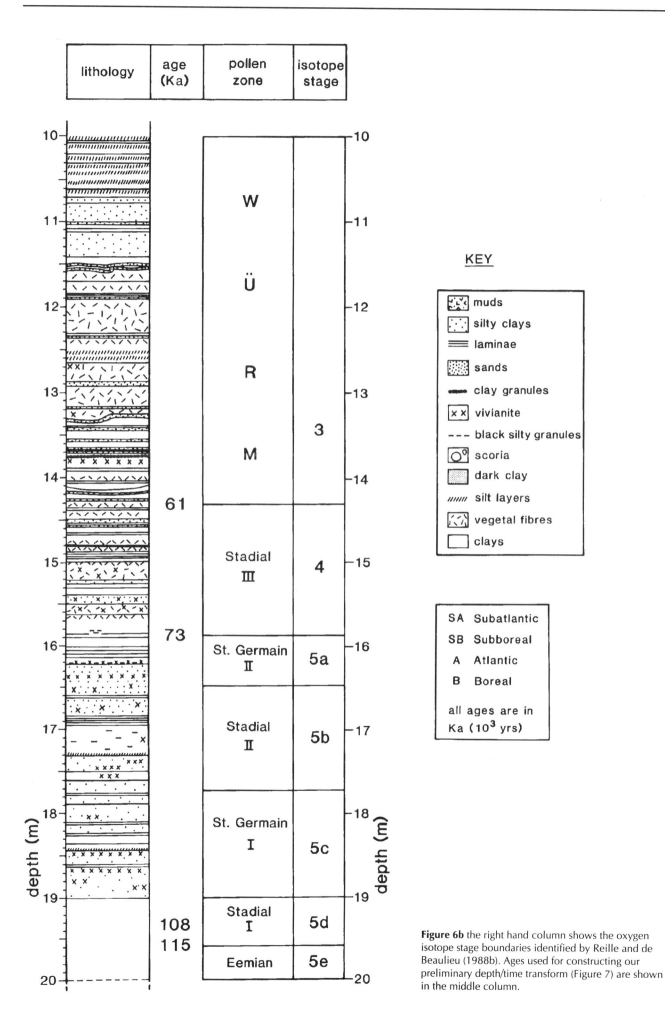

KEY

	muds
	silty clays
	laminae
	sands
	clay granules
x x	vivianite
- - -	black silty granules
O⁰	scoria
	dark clay
	silt layers
	vegetal fibres
	clays

SA   Subatlantic
SB   Subboreal
 A   Atlantic
 B   Boreal

all ages are in
Ka (10³ yrs)

Figure 6b the right hand column shows the oxygen isotope stage boundaries identified by Reille and de Beaulieu (1988b). Ages used for constructing our preliminary depth/time transform (Figure 7) are shown in the middle column.

Five more ages were determined on core B63 from the 12m Mackereth core series using the Accelerator Mass Spectrometer (AMS) facility at Tucson.

From the GEOMAAR series, four accelerator radiocarbon ages were measured on core D and five on core F with the HEMS at Oxford.

Dating by correlation with the oxygen isotope time-scale

The lower parts of the cores have been dated by correlating the pollen diagrams with the marine oxygen isotope stages. Figure 6b shows the ages of the identified stage boundaries taken from Hays, Imbrie and Shackleton (1976) alongside a lithological column drawn up by Pailles (1989).

Reille and de Beaulieu (1988b) correlated the Lac du Bouchet pollen profiles measured between 14m and 20m down core D with the previously studied pollen profiles for Le Grande Pile (Woillard, 1978) and for Les Echets (de Beaulieu and Reille, 1984). They identified, going from young to old, two warm stages with the St Germain II and I interstadials and a third warm stage with the top of the Eemian. Following the approach of Woillard and Mook (1982) these three interstadials were correlated with oxygen isotope stages 5a, 5c, and 5e. Two stadials separating them, labelled Stadial II and Stadial I in Figure 6b, also named Melissey II and I by Guiot *et al.* (1989), are correlated with oxygen isotope stages 5b and 5d.

Hays, Imbrie and Shackleton (1976) estimated the age of the stage 3/4 boundary (point 40, Table 1) at 61 ka and the 4/5a boundary (point 41, Table 1) at 73 ka. These ages are about 2 to 3 ka younger than those estimated by Shackleton and Opdyke (1973) by assuming a uniform sedimentation rate down core V28–238 where these stages were first recognized. The V28–238 ages were used by Pailles (1989). In addition, point 42 at 108 ka corresponds to the middle of the Melissey I stadial (isotope stage 5d) and point 43 at 115 ka to the 5d/5e boundary.

Construction of a depth to time transform

The age control data of Table 1 are plotted in Figure 7. Crosses represent pollen data, squares represent conventional radiocarbon ages from Gif, diamonds AMS ages from Oxford, circles AMS ages from Tucson, erect triangles AMS ages from Oxford on cores D and F, and inverted triangles represent ages inferred from correlation of pollen analyses on the older parts of core D with the marine oxygen isotope stages. The scatter of the radiocarbon ages is most likely due to the systematic errors associated with the 'hard water' effect and recycled organic material inherent in the application of the radiocarbon method to the dating of sediments, and also to the low organic carbon content which is only of the order of a few per cent.

Transforms spanning the end of the last glacial cycle from the Holocene to the late Wurm were developed by Creer *et al.* (1986) and Bonifay *et al.* (1987) for the 6m core series (core B5) and then by Smith and Creer (1986) for the 9m core series (core B49).

Various options have been considered for constructing a depth/time transform function through the pre-Holocene data points spanning the bulk of the last glacial cycle. These included a single straight line regression fit, several straight line segments with discontinuities of slope and cubic spline fits.

A first transform was developed by N. Thouveny. It incorporated six straight line segments constructed after rejecting age control points with low carbon content or large error bars (Thouveny, 1990; and Thouveny *et al.*, 1990).

An alternative set of transforms using smooth, continuous cubic spline transform functions were constructed to prepare the time-series for spectral analysis (Creer *et al.*, 1990). These incorporate additionally the set of accelerator dates from cores D and F. Points used in the computations are indicated by large symbols and those rejected by small symbols. The latter are: points 14 and 17 which were superceded by points 15 and 18, point 29 of which point 30 is a duplicate, point 26 and point 39 because of a very low organic carbon content (0.34%). This last age in particular stands out as being grossly erratic, and it is noted that the sample originates from a zone with anomalous magnetic anisotropy (Blunk, 1989, 1991). All the spline curves calculated with from 1 to 4 knots were simply concave, reflecting the progressive increase of compaction with increasing overburden. Curves with two changes in slope resulted for spline curves with 5, 6 and 7 equally spaced knots. The curve with 6 knots was chosen to transform to a time-scale.

The smooth transform curve shown in Figure 7 should be regarded as provisional because it neglects the influence of changes in sediment accumulation rate that must have occurred considering the dry conditions indicated for *Facies C* and *D* and more humid conditions for *Facies F* by clay mineral analyses. But the age controls currently available do not permit the construction of an unambiguous transform reflecting these shorter time-scale variations. Although it was found possible to construct a transform function with better resolution for the depth range between 1.07 and 9.33m, where the age control points are more densely concentrated, by fitting cubic spline curves constrained by a larger number of knots, curves with more than 8 knots could not be fitted to the whole data set because of the large spacing of the control points older than ~50 ka.

Clearly a prime objective of future work must be the construction of a more complex transform function.

Discussion

Maar lake sediment cores can provide long quasi-continuous sequences of sediments recording continental palaeo-environments. The importance of using *sequences* of samples rather than isolated samples cannot be overstressed because at least the order of the true ages is known and some, at least rough, estimate of the absolute ages can be made from intelligent guesses about sedimentation rates so identifying systematic errors which though large could remain undetected in the study of isolated samples.

The oldest of the radiocarbon ages obtained on the Lac du Bouchet cores are at the extreme limit of the capability of the radiocarbon method, given (i) current sample preparation methods, (ii) unavoidable atmospheric contamination of the apparatus and (iii) detector sensitivities. At the Tucson laboratory (in 1989) the 'chemical blank' equalled 4 μg of modern carbon yielding 0.42% modern for an infinitely old sample and the s.d. on repeat measurements was ±0.12% modern or 1.14μg of modern carbon: it has since been improved. The oldest age attainable by the Tucson AMS facility is currently about 46 ka: state of the art β-counting can yield much older meaningful dates without ^{14}C enrichment, ca. 60 ka, but requires large samples (Damon and Jull, personal communication).

The most intractable difficulty in dating however stems from systematic errors. So far as radiocarbon dating of sediments is concerned, these originate from the deposition of older these

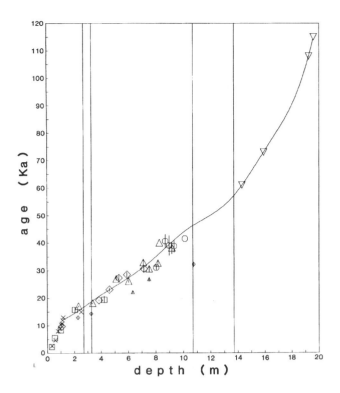

Figure 7 Plot of age control data taken from Table 1. The larger symbols represent those points used to compute the fitted curve (a cubic spline with 6 equally spaced knots - see text). The smaller symbols identify the rejected data points. The error bars for the radiocarbon ages relate only to analytical errors. Crosses indicate pollen data, squares conventional 14C ages from Gif, diamonds AMS ages from Oxford, circles HEMS ages from Tucson, erect triangles a second set of Oxford ages and inverted triangles ages inferred by correlation of % arboreal pollen with the oxygen isotope time-scale. The vertical lines indicate zones of anomalous magnetic anisotropy, from 2.7 to 3.3m and from 10.7 to 13.7m (see text).

components (*e.g.* redeposited soils and micro-fossils), bioturbation (which can be ruled out for large differences) and atmospheric ^{14}C fluctuations (which is not likely to cause a time-scale distortion of more than ~350 years during a millenium). The first of these problems can be attacked by separating out components such as birch fruits that can be identified and are unlikely to have been deposited after a long soil residence time. The other first-order difficulty is outside direct experimental control: it is the discrepancy between radiocarbon years from calendar years. The difference between the two time-scales has been measured for much of the Holocene by counting tree rings. In some maar lake sediments (Meerfelder Maar and Holzmaar in the Eifel region of Germany) Zolitschka and Negendank (1987) and Zolitschka (1989) have demonstrated that this type of approach can be extended through late glacial time at least, by counting micro-varves.

A combination of radiocarbon age control with the isotope time-scale clearly holds a strong potential for extending the range of calibration of radiocarbon to calendar years well beyond the tree ring range, though not with a year to year resolution. It is possible that the shallowing of the 6 knot transform curve shown in Figure 7 between about 10 and 14m depth may originate from a change-over from radiocarbon years to calendar years implicit in our time axis. This could be

interpreted as suggesting that radiocarbon year ages are around 5 ka *older* than calendar year ages at 40 ka to 60 ka. But it would be premature to accept this conclusion as firm because magnetic anisotropy studies (Blunk, 1989) indicate the possibility of a minor slumping episode between ~10.7m and ~13.7m depth, marked by vertical lines in Figure 7 (also see section 2).

It is noted that a systematic difference has been reported between ^{14}C ages and U-Th ages of a series of coral samples spanning the last 30 ka (Bard *et al.*, 1990). These authors argued that since the U-Th ages agreed with *calibrated* ^{14}C ages through the last 9 ka they could, in principle, be used to calibrate older radiocarbon ages. But their results indicate that the ^{14}C ages are ~3.5 ka *younger* than corresponding U-Th ages at ~20 ka BP, *i.e.* the discrepancy is in the opposite sense from that inferred in the last paragraph.

Future work

As of now a considerable improvement could be made to our provisional time-scale by further inspection of the existing pollen data to try to identify warm and cold phases within stages 2 and 3 and then using the stacked isotope stratigraphy of Prell *et al.*, (1986) dated by 'Specmap' (Imbrie *et al.*, 1984). More detailed palynological work (*i.e.* on more closely spaced samples), should allow more detailed correlation with the oxygen isotope time-scale. But this would be very time consuming.

Another possibility is the use of magnetic properties such as susceptibility for providing rapid initial age control because, as shown in Figure 8, there is a broad overall inverse relationship between magnetic susceptibility and the percentage of arboreal pollen. Susceptibility is strong in sediment deposited during stadials and low in those with high organic content deposited during interstadials, especially during the Holocene. Thus it seems that susceptibility can provide a good proxy record of climatic change. In fact it is likely that a finer scale correlation is masked by some discrepancies between the time-scales constructed independently in the pollen and palaeomagnetic laboratories, because it is possible that steps taken to removehave not been completely successful.

Acknowledgements

The palynological work was carried out by the group at the Laboratoire de Botanique historique et Palynologie, Faculté des Sciences et Techniques Saint-Jérôme, Marseille (A. Pons, J-L de Beaulieu and M. Reille). The sedimentological studies were made by E. Truze at the Laboratoire de Géologie du Quaternaire, Faculté des Sciences de Luminy, Marseille. The GEOMAAR project was funded by a grant from the Stimulation Action Programme of the European Community (ST2J-128) under the direction of E. Bonifay (Marseille), together with J. Negendank (Trier) and K.M. Creer (Edinburgh). The accelerator radiocarbon age determinations were made by R. Hedges (Oxford) and P. Damon and A. Jull (Tucson). A Pike (Edinburgh) and H. Usinger (Kiel) and R. Hansen (Trier) directed the technical operations of the Mackereth and Livingstone coring respectively. Thanks are due to all these named scientists and to many others who have helped with the field and laboratory work over the last 8 years.

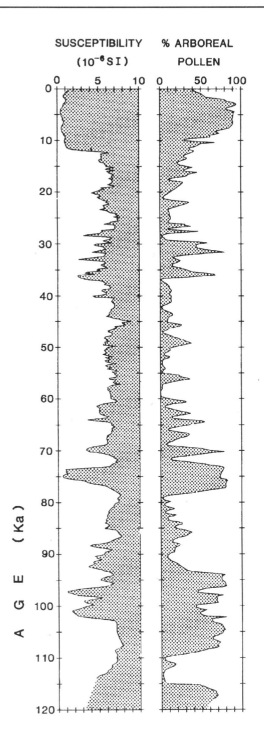

SUSCEPTIBILITY   % ARBOREAL

(10⁻⁶ S I)   POLLEN

Figure 8 Comparison of magnetic susceptibility [plotted on left] and the proportion of arboreal pollen [plotted on right] suggesting that the former provides a proxy record of climatic change.

References

BARD, E., HAMELIN, B., FAIRBANKS, R.G. and ZINDLER, A. (1990). Calibration of the ¹⁴C timescale over the past 30000 years using mass spectrometric U-Th ages from Barbados corals. *Nature*, 345, 405-410.

BONIFAY, E. and TRUZE, E. (1987). Dynamique sedimentaire et evolution des lacs de maars: l'exemple du velay – Travaux francais en Palaeolimnologie; Actes du Colloque du Puy en Velay, 4-6 Oct 1985, organisé par E. Bonifay et J. Mergoil, *Documents du Centre d'Etudes et de Recherche sur les Lacs, Anciennes Lacs et Tourbières (C.E.R.L.A.T.)*, Memoire Nr 1, 29-64.

BLUNK, I. (1989). Magnetic susceptibility anisotropy and deformation in Quaternary lake sediments. *Zeitschrift Deutschen Geologischen Gesellschaft 140*, 393-403.

BLUNK, I. (1991). Magnetic fabric studies of sediments from Lac du Bouchet, Actes du Colloque du Pay en Velay, 3-4 May 1988, organisé par E. Bonifay, *Documents du Centre d'Etudes et de Recherche sur les Lacs, Anciennes Lacs et Tourbieres (C.E.R.L.A.T.)*, Memoire Nr 2, – in press.

BONIFAY, E., CREER, K.M., DE BEAULIEU, J.L., CASTA, L., DELIBRIAS, G., PERINET, G., PONS, A., REILLE, M., SERVANT, S., SMITH, G., THOUVENY, N., TRUZE, E. and TUCHOLKA, P. (1987). Study of the Holocene and Late Wurmian sediments of Lac du Bouchet, (Haute Loire, France): First Results. *In Climate – History, Periodicity and Predictability*, (eds. Rampino, M.R., Sanders, J.E., Newman, W.S. and Konigsson, L.K.), van Nostrand Reinhold Co., New York, pp90 -116.

CREER, K.M., (1989). The Lac du Bouchet palaeomagnetic record: its reliability and some inferences about the character of geomagnetic secular variations through the last 50000 years. *In*: Lowes, F.J., Collinson, D.W., Parry, J.H., Runcorn, S.K., Tozer, D.C. and Soward, A. (Eds): *Geomagnetism and Palaeomagnetism;* Nato *ASI Series* C 262, Kluwer Academic Publishers, Dordrecht pp71-89.

CREER, K.M., SMITH, G., TUCHOLKA, P., BONIFAY, E., THOUVENY, N. and TRUZE, E. (1986). A preliminary palaeomagnetic study of the Holocene and Late Glacial sediments of Lac du Bouchet (Haute Loire), France. *Geophysical Journal Royal astronomical Society*, 86, 943-964.

CREER, K.M., THOUVENY, N. and BLUNK, I. (1990). Climatic and geomagnetic influences on the Lac du Bouchet palaeomagnetic SV record through the last 110000 years. *Physics of Earth and planetary Interiors*, 64, 314-341.

CREER, K.M., THOUVENY, N., BLUNK, I., SMITH G. and TURNER, G. (1991). Extension of the palaeomagnetic record from Lac du Bouchet, Haute Loire, France to about 50000 years before present. Actes du Colloque du Puy en Velay, 3-4 May 1988, organisé par E. Bonifay, *Documents du Centre d'Etudes et de Recherche sur les Lacs, Anciennes Lacs et Tourbières (C.E.R.L.A.T.)*, Memoire Nr2 – in press.

de BEAULIEU, J.-L., and REILLE, M. (1984). A long upper Pleistocene pollen record from Les Echets near Lyon, France. *Boreas*, 13, 111-132.

de BEAULIEU, J.-L. and REILLE, M. (1987). Histoire de la vegetation Wurmienne et Holocène du Velay occidental (Massif Central, France): analyse pollinique comparée de trois sondages du Lac du Bouchet – Travaux francais en Paleolimnologie; Actes du Colloque du Puy en Velay, 4-6 Oct 1985, organisé par E. Bonifay et J. Mergoil, . *Documents du Centre d'Etudes et de Recherche sur les Lacs, Anciennes Lacs et Tourbières (C.E.R.L.A.T.)*, Memoire Nr 1, 113 - 132.

GUIOT, J., PONS, A., de BEAULIEU, J.-L. and REILLE, M. (1989). A 140000 year continental climate reconstruction from two European pollen records. *Nature*, 338, 309-313.

HAAS, H.-C., (1989). Sedimentologische und schwermineralogische Untersuchungen an ausgewahlten Sedimenttypen des Lac du Bouchet. *Diplon Arbeit im Fachbereich III. Geographie und geowissenschaften.* Universitat Trier.

HAAS, H.-C. and NEGENDANK, J.F.W., (1989). Sedimentologie und Schwermineralanalyse an ausge-wahlten sedimenttypen des Lac du Bouchet (Massif Central, Frankreich). Actes du Colloque du Puy en Velay, 3-4 May 1988, organisé par E. Bonifay, *Documents du Centre d'Etudes et de Recherches sur les Lacs, Anciennes Lacs et Tourbières (C.E.R.L.A.T.)*, Memoire Nr 2 – in press.

HAYS, J.D., IMBRIE, J. and SHACKLETON, N.J., (1976). Variations in the Earth's orbit, pacemaker of the ice ages. *Science*, 194, 1121-1132.

IMBRIE, J., HAYS, J.D., MARTINSON, D.G., MCINTYRE, A., MIX, A.C. , MORLEY, J.J., PISIAS, N.G., PRELL, W.L. and SHACKLETON, N.J. (1984). The Orbital Theory of Pleistocene Climate: Support from a Revised Chronology of the Marine $\varnothing^{18}O$ Record. In *Milankovitch and Climate, Part I* (eds Berger, A.L., Imbrie, J., Hays, J., Kukla, G. and Saltzman, B). *N.A.T.O. ASI Series C: Mathematical and Physical Sciences*, Vol 126, D. Reidel Publishing Co., Dordrecht, Holland, pp 269-305.

KENT, D.V., (1982). Apparent correlation of palaeomagnetic intensity and climate records in deep sea sediments. *Nature*, 299, 538-539.

KUKLA, G., HELLER, H., LIU, X.M., XU, T.C., LIU, T.S. & AN, Z.S. (1988). Pleistocene climates in China dated by magnetic susceptibility. *Geology*, 16, 811-814.

LOWE, J.J., LOWE, S., FOWLER, A.J., HEDGES, R.E.M. & AUSTIN, T.J.F. (1988). Comparison of accelerator and radiocarbon measurements obtained from Late Devensian Lateglacial lake sediments from Llyn Gwernan, South Wales, UK., *Boreas*, 17, 355-369.

MACKERETH, F.J.H. (1958). A portable core sampler for lake deposits. *Limnology and Oceanography*, 3, 181 - 191.

PAILLES, C. (1989). Les Diatomées du Lac de Maar du Bouchet (Massif Central, France). Reconstruction des Paleoenvironnements au cours des 120 derniers millenaires. *Thèse de doctorat. Université d'Aix-Marseille II.* 271pp.

PRELL, W.L., IMBRIE, J., MARTINSON, D., MORLEY, J.J., PISIAS, N.G., SHACKLETON, N.J. and STREETER, H.F. (1986). Graphic correlation of Oxygen Isotope Stratigraphy application to the Late Quaternary. *Palaeoceanography*, 1, 137-162.

REILLE, M. and de BEAULIEU, J.L., (1988a). History of the Wurm and Holocene vegetation in the western Velay (Massif Central, France). A comparison of pollen analysis from three corings at Lac du Bouchet. *Review Palaeobotany Palynology*, 54,233-248.

REILLE, M. and de BEAULIEU, J-L. (1988b). La fin de L'Eemian et les interstades du Prewurm mis pour la première fois en évidence dans le massif central francais par l'analyse pollinique. *Comptes Rendues Academie des Sciences, Paris*, 306, 1205-1210.

SHACKLETON, N.J., and OPDYKE, N.D. (1973). Oxygen isotope and palaeomagnetic stratigraphy of equatorial Pacific core V28-238: Oxygen isotope temperatures and ice volumes on a 10^5 and 10^6 year scale. *Quaternary Research*, 3, 39-55.

SMITH, G. (1985). Late glacial palaeomagnetic secular variations from France. *PhD Thesis, University of Edinburgh*, 115pp.

SMITH, G. AND CREER, K.M. (1986). Analysis of geomagnetic secular variations 1000 - 30000 years b.p., Lac du Bouchet, France. *Physics of Earth and planetary Interiors*, 44, 1 - 14.

THOUVENY, N. (1983). Etude paleomagnetique de formations du Plio-Pleistocene et de l'Holocène du Massif central et de ces abords: contribution a la géologie du Quaternaire. *Thèse de 3e cycle, Université d'Aix-Marseille II.*

THOUVENY, N. (1990). Variations du champ magnétique terrestre dans le dernier cycle climatique (0 - 120000 ans BP): Memoire d'Habilitation à dirigir les Rescherches, *Université d'Aix-Marseille*, 192pp.

THOUVENY, N., CREER, K.M. and BLUNK, I. (1990). Extension of Lac du Bouchet palaeomagnetic record over the last 120000 years, *Earth and planetary Science Letters*, 97, 140-161.

THOUVENY, N., CREER, K.M., and BLUNK, I. (1991). Initial palaeomagnetic results for cores A - D from Lac du Bouchet, 0 - 125 Myr BP. Actes du Colloque du Puy en Velay, 3.4 May 1988, organisé par E. Bonifay, *Documents du Centre d'Etudes et de Recherche sur les Lacs, Anciennes Lacs et Tourbières (C.E.R.L.A.T.)*, Memoire Nr 2, – in press.

TRUZE, E. (1988). Etude préliminaire de la sedimentation dans les lacs de maars du Deves. Le Lac du Bouchet. *Marseille: Memoire de Diplome d'études approfondies*, 254pp.

WOILLARD, G. (1978). Grande Pile Peat Bog: a continuous pollen record for the last 140000 years. *Quaternary Research*, 9, 1-21.

WOILLARD, G. and MOOK, W.G., (1982). Carbon 14 dates at La Grande Pile: correlation of land and sea chronologies. *Science*, 215, 159-161.

ZOLITSCHKA, B. (1989). Jahreszeitlich geschichtete Seesedimente aus dem Holzmaar und dem Meerfelder Maar (Westeifel), *Zeitschrift Deutschen Geologischen Gesellschaft*, 140, 25-33.

ZOLITSCHKA, B. and NEGENDANK, J.F.W. (1987). Annually laminated lake sediments of the Meerfeld Maar / Eifel (FRG) and their implications on the late Pleistocene and Holocene stratigraphy. *Terra Cognita*, 7, 220-221.

QUATERNARY RESEARCH ASSOCIATION

The **Quaternary Research Association** is an organisation comprising archaeologists, botanists, civil engineers, geographers, geologists, soil scientists, zoologists and others interested in research into the problems of the Quaternary. Most members reside in Great Britain, but membership also extends to most European countries, North America, Africa and Australia. Current membership stands at circa 1000. Membership is open to all interested in the objectives of the **Association.** The annual subscription for ordinary members is £10.00 and is due on January 1st for each calendar year. Reduced rates apply for students and unwaged members.

The main meetings of the **Association** are the Annual Field Meeting, usually lasting 3 or 4 days, held in April, and a 1 or 2 day Discussion Meeting held at the beginning of January. Additionally, Short Field Meetings may be held in May or September and occasionally these visit overseas locations. Short Study Courses on the techniques used in Quaternary work are also held. The publications of the **Association** are the *Quaternary Newsletter* issued with the **Association's** *Circular* in February, June and November; the *Journal of Quaternary Science* published in association with Wiley, and with four issues a year; the Field Guides Series, the Technical Guides Series and *Quaternary Proceedings.*

The **Association** is run by an executive committee elected at an annual general meeting held during the April Field Meeting. The current officers of the **Association** are:

President:
Professor G.S. Boulton, Grant Institute of Geology, University of Edinburgh, West Mains Road, Edinburgh EH9 3JW.

Vice-President:
Professor W. A. Watts, Provost's House, Trinity College, Dublin 2, Ireland.

Secretary
Dr. M.J.C. Walker, Department of Geography, St. David's University College, University of Wales, Lampeter, Dyfed SA48 7ED, Wales.

Assistant Secretary (Publications):
Dr. D.R. Bridgland, 41 Geneva Road, Darlington, C. Durham, DL1 4NE.

Treasurer:
Dr. C.A. Whiteman, Department of Geography, Royal Holloway and Bedford New College, University of London, Egham, Surrey TW20 OEX.

Editor, Quaternary Newsletter:
Dr. B.J. Taylor, British Geological Survey, Keyworth, Nottinghamshire NG12 5GG.

Editor, Journal of Quaternary Science:
Dr. P.L Gibbard, Subdepartment of Quaternary Research, Botany School, Downing Street, Cambridge CB2 3EA.

All questions regarding membership are dealt with by the Secretary, the **Association's** publications are sold by the Assistant Secretary (Publications) and all subscription matters are dealt with by the Treasurer.

NOTE FOR PROSPECTIVE EDITORS

Quaternary Proceedings is an occasional publications series established to report the proceedings of important meetings held under the aegis of the *Quaternary Research Association* or organised jointly by the *QRA* and other scientific organisations. Frequency of publication will depend upon the receipt of proposals of reports of meetings which provide the basis for a coherent proceedings volume. Proposals should be submitted in the first instance to:

Series Editor, Quaternary Proceedings:
Professor J.John Lowe, Department of Geography,
Royal Holloway , University of London,
Egham, Surrey TW20 0EX, U.K.

The following guidelines must be adhered to in the production of volumes for the *Quaternary Proceedings* series:

1. All issues must conform with previous issues of *Quaternary Proceedings* in terms of quality of materials, type-set and printing (of both cover and text pages), arrangement of headings, size of lettering, general design and the use of the *QRA* logo. Reference can be made to the most recent issue of *QP* for guidance, and further particulars are available from the **Series Editor**.

2. Scientific Conventions and preparation of text and figures follow the 'Instructions for Authors' for papers published by the *Journal of Quaternary Science*. These are summarised on the inside back cover of issues of that journal.

3. All proceedings proposals must obtain the prior approval of the *PUBLICATIONS SUB-COMMITTEE* of the *Quaternary Research Association*. Mock cover designs and text pages should be submitted for approval along with other general information of publication proposals.

4. Prospective editors should also supply details of arrangements for type-setting and printing and of production costs of the volume for prior approval.

5. All volumes are produced under the sponsorship of the *QRA* and all profits reside with the *QRA*.

6. Copyright for *Quaternary Proceedings* resides with the *QRA*.

7. An ISSN number exists for *QP* (ISSN 0963-1895) and all volumes in the series must conform with the specifications to which this refers. For advice on this matter, consult the Assistant Secretary (Publications) of the *QRA*.